Wilfried Elmenreich   Falko Dressler
Vittorio Loreto (Eds.)

# Self-Organizing Systems

7th IFIP TC 6 International Workshop, IWSOS 2013
Palma de Mallorca, Spain, May 9-10, 2013
Revised Selected Papers

 Springer

Volume Editors

Wilfried Elmenreich
Alpen-Adria-University of Klagenfurt
Institute of Networked and Embedded Systems
Klagenfurt, Austria
E-mail: wilfried.elmenreich@aau.at

Falko Dressler
University of Innsbruck
Institute of Computer Science
Innsbruck, Austria
E-mail: dressler@ccs-labs.org

Vittorio Loreto
Sapienza University of Rome
Physics Department
Rome, Italy
E-mail: vittorio.loreto@roma1.infn.it

ISSN 0302-9743      e-ISSN 1611-3349
ISBN 978-3-642-54139-1      e-ISBN 978-3-642-54140-7
DOI 10.1007/978-3-642-54140-7
Springer Heidelberg New York Dordrecht London

Library of Congress Control Number: 2013957941

CR Subject Classification (1998): C.2, D.4.4, C.2.4, C.4, H.3, I.2.11

LNCS Sublibrary: SL 5 – Computer Communication Networks and Telecommunications

*Typesetting:* Camera-ready by author, data conversion by Scientific Publishing Services, Chennai, India

Printed on acid-free paper

Springer is part of Springer Science+Business Media (www.springer.com)

# Lecture Notes in Computer Science     8221

Commenced Publication in 1973
Founding and Former Series Editors:
Gerhard Goos, Juris Hartmanis, and Jan van Leeuwen

# Lecture Notes in Computer Science 8221

Commenced Publication in 1973
Founding and Former Series Editors:
Gerhard Goos, Juris Hartmanis, and Jan van Leeuwen

# Preface

This book contains the research articles that were presented at the International Workshop on Self-Organizing Systems (IWSOS) held in Palma de Mallorca, Spain, in May 2013. This was the seventh workshop in a series of multidisciplinary events dedicated to self-organization in networks and networked systems, including techno-social systems.

Self-organization relates the behavior of the individual components (the microscopic level) to the resulting networked structure and functionality of the overall system (the macroscopic level), where simple interactions at the microscopic level may already give rise to complex, adaptive, and robust behavior at the macroscopic level. On the other hand, designing self-organizing systems comes with new challenges such as their controllability, engineering, testing, and monitoring. Therefore, the IWSOS workshop series provides a highly multidisciplinary and innovative forum for researchers from different areas to exchange ideas and advance the field of self-organizing systems. The growing scale, complexity, and dynamics of (future) networked systems have been driving research from centralized solutions to self-organized networked systems. The applicability of well-known self-organizing techniques to specific networks and networked systems is being investigated, as well as adaptations and novel approaches inspired by cooperation in nature. Models originating from areas such as control theory, complex systems research, evolutionary dynamics, sociology, and game theory are increasingly applied to complex networks to analyze their behavior, robustness, and controlability. Self-organization principles not only apply to the Internet and computer networks but also to a variety of other complex networks, like transportation networks, telephony networks, smart electricity grids, financial networks, social networks, and biological networks. "Network science" and "complex networks theory" constitute new research areas that provide additional insights into self-organizing systems.

This year, we received 35 paper submissions. On the basis of the recommendations of the Technical Program Committee and external expert reviewers, we accepted 11 full papers and nine short papers. Most full papers were reviewed by four experts, and all papers received at least three reviews. The workshop featured a keynote lecture by Alessandro Vespignani on modeling and forecast of socio-technical systems, as well as two round tables on *Techno-Social Systems* and *Future Control Challenges for Smart Grids*. The papers presented in the talks addressed the topics of data dissemination, energy systems and smart grids, evolutionary algorithms, social systems, transportation networks, and wireless sensor networks.

We are grateful to all TPC members and additional reviewers who provided thorough reviews that made the selection of the papers possible. Special thanks go to our IWSOS 2013 general chairs, Maxi San Miguel and Hermann de Meer, for their outstanding support in all the phases of the workshop organization. Additionally, thanks goes to our treasurer Bernhard Plattner, Pere Colet, who managed the local organization, and the publicity chairs, Karin Anna Hummel and Carlos Gershenson. Finally, we want to thank the authors for their submissions and contributions to the technical program.

November 2013

Wilfried Elmenreich
Falko Dressler
Vittorio Loreto

# Organization

IWSOS 2013, the 7th International Workshop on Self-organizing Systems, was organized by the Institute for Cross-Disciplinary Physics and Complex Systems, a joint research Institute of the University of the Balearic Islands (UIB) and the Spanish National Research Council (CSIC) on the University of Balearic Islands Campus, Palma de Mallorca, Spain, May 9–10, 2013.

## Steering Committee

| | |
|---|---|
| Hermann de Meer | University of Passau, Germany |
| David Hutchinson | Lancaster University, UK |
| Bernhard Plattner | ETH Zurich, Switzerland |
| James Sterbenz | University of Kansas, USA |
| Randy Katz | UC Berkeley, USA |
| Georg Carle | TU Munich, Germany (IFIP TC6 Representative) |
| Karin Anna Hummel | ETH Zurich, Switzerland |
| Shlomo Havlin | Bar-Ilan University, Israel |

## General Chairs

| | |
|---|---|
| Maxi San Miguel | IFISC (CSIC-University of the Balearic Islands), Spain |
| Hermann de Meer | University of Passau, Germany |

## Program Chairs

| | |
|---|---|
| Falko Dressler | University of Innsbruck, Austria |
| Vittorio Loreto | Sapienza University of Rome, Italy |

## Publicity Chairs

| | |
|---|---|
| Karin Anna Hummel | ETH Zurich, Switzerland |
| Carlos Gershenson | Universidad Nacional Autonoma de Mexico, Mexico |

## Publication Chair

Wilfried Elmenreich                University of Klagenfurt and Lakeside Labs,
                                   Austria

## Poster Chair

Wilfried Elmenreich                University of Klagenfurt and Lakeside Labs,
                                   Austria

## Treasurer

Bernhard Plattner                  ETH Zurich, Switzerland

## Local Organization

Pere Colet                         IFISC (CSIC-University of the Balearic
                                   Islands), Spain

## Technical Program Committee

Karl Aberer                        EPFL, Switzerland
Andrea Baronchelli                 Northeastern University, USA
Alain Barrat                       Centré de Physique Théorique, France
Marc Barthelemy                    Institut de Physique Théorique, France
Christian Bettstetter               University of Klagenfurt, Austria
Raffaele Bruno                     Consiglio Nazionale delle Ricerche (CNR), Italy
Claudio Castellano                 CNR-ISC Rome, Italy
Ciro Cattuto                       ISI Foundation Turin, Italy
Hermann de Meer                    University of Passau, Germany
Albert Diaz-Guilera                University of Barcelona, Spain
Falko Dressler                     University of Innsbruck, Austria
Alois Ferscha                      Johannes Kepler University of Linz, Austria
Andreas Fischer                    University of Passau, Germany
Santo Fortunato                    Aalto University, Finland
Carlos Gershenson                  Universidad Nacional Autonoma de Mexico,
                                   Mexico
Salima Hassas                      University of Lyon 1, France
Boudewijn Haverkort                University of Twente, The Netherlands
Poul Heegaard                      Norwegian University of Science and
                                   Technology, Norway
Tom Holvoet                        Katholieke Universiteit Leuven, Belgium
Karin Anna Hummel                  ETH Zurich, Switzerland
Sebastian Lehnhoff                 OFFIS Institute for Information Technology,
                                   Germany

| | |
|---|---|
| Vittorio Loreto | Sapienza University of Rome, Italy |
| Hein Meling | University of Stavanger, Norway |
| Yamir Moreno | BIFI, University of Zaragoza, Spain |
| Mirco Musolesi | University of Birmingham, UK |
| Dimitri Papadimitriou | Alcatel-Lucent Bell, Belgium |
| Christian Prehofer | Fraunhofer ESK, Germany |
| Jose Ramasco | IFISC (CSIS-UIB), Spain |
| Andreas Riener | Johannes Kepler University of Linz, Austria |
| Kave Salamatian | Université De SavoieMarc Barthelemy, France |
| Maxi San Miguel | IFISC (CSIC-UIB), Spain |
| Hiroki Sayama | Binghamton University, USA |
| Paul Smith | Austrian Institute of Technology, Austria |
| Bosiljka Tadic | Jozef Stefan Institute, Slovenia |
| Dirk Trossen | University of Cambridge, UK |

# Table of Contents

# Automated
# Trading for Smart Grids: Can It Work?

Barry Laffoy[1,2], Saraansh Dave[1,3], and Mahesh Sooriyabandara[1]

[1] Toshiba Research Europe Ltd., Telecommunications Research Laboratory
[2] Department of Computer Science, University of Bristol
[3] Industrial Doctorate Centre in Systems, University of Bristol
{saraansh.dave,mahesh.sooriyabandara}@toshiba-trel.com

**Abstract.** This paper applies basic economic principles which have been developed in financial markets to a future smart grid scenario. Our method allows for autonomous bidding for electricity units to create an emerging market price for electricity. We start with replicating the popular Zero-Intelligence-Plus algorithm and setting it in a electricity supplier-consumer scenario. We identify significant weaknesses of applying this in an electricity market especially when intermittent sources of energy are present or when the supplier to consumer ratio is very small. A new algorithm (ZIP-260) is proposed which includes a measure of fairness based on minimising the deviation across all un-matched demand for a given period. This approach means that no consumer in the system is constantly experiencing an electricity supply deficit. We show and explain how market conditions can lead to collective bargaining of consumers and monopolistic behaviour of suppliers and conclude with observations on automated trading for smart grids.

## 1 Introduction

Financial markets have been used in recent history as a robust method for matching buyers and sellers. Since the late 1990's computers have been increasingly used as intelligent agents to trade on behalf of human operators at a much higher frequency. High frequency trading has been bolted onto pre-existing system regulation and this may enhance the effects where billions of dollars have been lost in a span of just 15 minutes [1] [2]. The root cause of some of these 'dips' such as the flash crash of May 2006 have not been unilaterally agreed on but have been classed as a systemic failure [1]. However, financial markets can be said to be a very efficient method for solving resource allocation problems in decentralised systems with multiple actors.

Smart grids carry the promise of real-time energy management, diagnostics and fault correction with the aim of integrating renewable energy sources and decarbonised transport solutions [3]. The smart grid is characterised as integrating information and communication technologies with the electricity network to create a responsive and adaptable network. The primary challenge with intermittent energy sources and real-time balancing is trying to match demand patterns to those of supply. In a future where decarbonisation of energy is paramount, the ability to respond to demand fluctuations will become more difficult. Demand

W. Elmenreich, F. Dressler, and V. Loreto (Eds.): IWSOS 2013, LNCS 8221, pp. 1–13, 2014.
© IFIP International Federation for Information Processing 2014

response [4] as a method of addressing this balancing issue can be described as reflecting the value of electricity at a point in time given the state of the system.

This view resembles the resource allocation problems found in real-time trading markets and we propose to investigate the application of algorithmic trading as an efficient mechanism to manage electricity demand for households with variable electrical supply. We apply the Zero-Intelligence-Plus (ZIP) algorithm [5], a well known automatic trading algorithm, to this problem scenario. Initially the algorithm is directly mapped onto a scenario including electricity suppliers and domestic consumers. After evaluating its performance, changes are made to allow for a more fair consumption distribution. Even after promoting fairness in our system we identify inefficient market behaviour and monopolistic behaviour when there is a supply surplus. This observation of different behaviour between deficit and surplus states of the market lead us to implementing a 2-state switching algorithm for the suppliers which accounts for this. We identify some key limitations of applying the ZIP algorithm and provide regulatory and aglorithmic requirements that are needed for real-time pricing of electricity when trying to balance supply and demand.

This paper is organized as follows; §2 covers relevant literature, §3 outlines the system model, §4 details the simulation scenarios used and results, §5 presents our analysis and §6 concludes on the work and outlines areas for further research.

## 2   Background Literature

Demand response has been classified as incentive based or priced based programs (IBP or PBP respectively). IBP includes more direct aspects such as direct load control or emergency demand response while PBP utilises more market oriented solutions such as real time pricing or time of use pricing. Time of use pricing has been shown to create a bias of consumer behaviour [6] which causes 'rebound' peaks to be seen during the off-peak tariff period. Clearly this method is limited in its approach and can have undesirable consequences when the solution is scaled up. Game-theoretic approaches show how the market can converge to a Nash equilibrium while improving the aggregate demand curve [7] but are somewhat limited as they usually assume a central coordinating agent (e.g. the utility) which makes it difficult to implement on a large scale. Real time pricing is more favourable towards the principle of using price as a signal of the value of electricity at that specific moment in time but there has been little work in the application and understanding of real time trading in such a system. The closest work in the smart grid domain applies transmission constraints and novel balancing mechanisms to trading agents which perform bidding and acceptance of electricity units [8].

More directly relevant work falls into the domain of agent-based computational economics [9] and continuous double auctions (CDA). CDA's are a market system where buy and sell prices are quoted continuously and asynchronously for commodities and is widely used around the world. Previous work on CDA's applied to automatic trading show that very simple behaviour of agents have high levels of allocative efficiency [10] and can be improved to also provide a high level of profit dispersion (an indicator of market equality) [5]. The Zero Intelligence

(ZI) traders' [10] private information only included the limit price for that trader, but by giving the agents an internal, mutable expected profit margin and memory of the last shout on the market, simple machine-learning techniques could give the traders more sophisticated behaviour [5]. The resulting Zero-Intelligence Plus (ZIP) agents would observe each shout on the market and remember if the last one was a bid or an offer, for how much, and whether or not it resulted in a trade. Based on this, the agent would update its internal profit margin and shout price.

This research also falls into the domain of current electricity markets which operate differently depending on country specific contexts. In the U.K. contracts are agreed between those demanding electricity and generators in half-hourly periods. When there is a difference between the contracted amount and generator supply the system operator (National Grid) is responsible for real-time balancing through bids and offers [11]. This basic model may have to be reviewed to allow for smart grid implementation such that the effect of wholesale prices can be observed by the consumer and thus influence demand patterns. Currently there is a buffer between real-time consumption and the price of electricity. The role of the system operator will still apply as the real-time management must be controlled to avoid grid failures.

Directly applying the ZIP alogrithm needs careful thought and experimentation as there are a few fundamental differences between electricity and financial markets. Financial markets have liquidity which assumes that a seller is able to sell their stock on the market if they wish to do so at a given price, additionally, a seller is able to withold stock in anticipation of better prices in the future. In an electrical market, it is not possible to withold a unit for a better price as electricity cannot be stored. Financial markets also have a large number of buyers and sellers which is in stark contrast to electricity generators and consumers where there is a much bigger difference. These differences make the application of ZIP to electricity markets non-trivial but also enhances the view that a real-time solution should be used in order to allow the system to respond at a similar rate, especially as more renewable energy is integrated into the grid.

## 3   System Model

This section outlines the scenario we use to test our algorithms and measure the behaviour of the system. Our simulation consists of 100 households and two electricity suppliers; fossil fuel (Supplier F) and wind power (Supplier W) suppliers. Supplier F is modelled as having constant power output reflecting a non-renewable energy source (e.g. coal or nuclear) while Supplier W has a variable power output modelled as following a Weibull distribution [12]. By including suppliers with different output profiles, we will be able to investigate whether the real-time (half-hourly) changes in supply affect the market price and whether the mechanism is efficient in allocating resources. The households are able to bid for units of electricity through the exchange (which could be organised by the utility) where Suppliers F and W are able to sell units of electricity. Households bid for a total number of units every half hour depending on their individual consumption profile whilst the suppliers are able to sell units every half hour depending on their plants output. Empirical smart meter data measured at half-hourly intervals in an Irish study [13] is used as the demand profile for households in the simulation. This also represents the future scenario of a decarbonised electricity supply where centrally generated base load capacity will support intermittent and decentralised sources. Both the heterogeneity in demand profiles and changes in aggregate supply and demand will create an environment with different market states which will test the algorithms' performance. A genetic algorithm was used to search the parameter space in order to find an optimal solution (we note that due to the nature of genetic algorithms this may well be a local optimum) where each genome expresses a phenotype in the form of a vector $[\mu, \beta, \lambda, \rho, \epsilon]$. The fitness function can be chosen according to the aim of the algorithm; a popular choice is Smith's $\alpha$ [14] which is a measure for how well the market converges to an equilibrium price. This measure did not promote fairness over the agents, especially in a situation where there is an overall deficit of supply. As a result we decide to use a measure which minimises the demand deficit variance across all the households (9). In the the remainder of this section, §3.1 introduces the original ZIP algorithm and §3.2 explains the limitations of directly applying this to electricity markets and the significant changes we made to the algorithm and §3.3 covers price and fitness functions.

## 3.1 The ZIP Algorithm

This sub-section is a recap and overview of the ZIP algorithm as first proposed by Cliff [5]. ZIP agents have an internal, mutable expected profit margin ($\mu_i$) which can be used with the current shout price ($\lambda_i$) to calculate the current shout price ($p_{i,t}$) as shown in (1). Based on this, it might alter it's internal profit margin. The rate at which an agent's profit margin converges to the observed shout price is determined by a simple version of the Widrow-Hoff delta rule for machine learning. This gives an update rule between the times of $t$ and $t+1$ as shown in (2), where $\Delta_{i,t}$ is defined in (3).

$$p_{i,t} = \lambda_i \times (1 + \mu_{i,t}) \tag{1}$$

$$\mu_{i,t+1} = (p_{i,t} + \Gamma_{i,t})/\lambda_i - 1 \tag{2}$$

$$\Delta_{i,t} = \beta_i \times (\tau_{i,t} - p_{i,t}) \tag{3}$$

$$\tau_{i,t} = q_{i,t} \times (1 + \rho) + \epsilon \tag{4}$$

$$\Gamma_{i,t+1} = \lambda_{i,t} \times \Gamma_{i,t} + (1 - \lambda_{i,t}) \times \Delta_{i,t} \tag{5}$$

Here, $\beta_i$ is the agent's learning rate, which must be added to the private information of each agent. $\tau_{i,t}$ is the target price, towards which the shout price is being moved and $\Gamma_{i,t}$ is the learning momentum. There is a distinction between the target price and the most recent market shout price, $q_{i,t}$ as otherwise the agents will no longer test the market if $p_{i,t} \approx q_{i,t}$, since this would lead to a value of $\Delta_{i,t}$ close to zero and thus $\mu_{i,t+1} \approx \mu_{i,t}$. The ZIP agents calculate the target value using a randomly generated relative multiplier ($\rho$) of the last shout price, plus a randomly generated absolute perturbation ($\epsilon$) shown in (4). Finally, the agents were given a learning momentum coefficient $\Gamma_{i,t}$, so that when updating prices, they could take into account the history of price updates. This dampens the impact of a series of price updates in alternate directions, and reinforces the effect of a series all in the same direction.

Thus the complete behaviour of a ZIP agent is defined by the initial profit assumption $\mu$ and the values of $\beta, \lambda, \rho,$ and $\epsilon$. One variant of the ZIP algorithm (ZIP60) [15] allowed agents to distinguish up to six different scenarios under which prices can be changed and create a rule-set for each. The ZIP algorithm also allows buyers and sellers to start with different assumptions hence embracing heterogeneity that is present in most trading environments.

## 3.2   ZIP-260

Unlike ZIP, we do not impose any scenario under which prices cannot be updated. To increase flexibility the agent considers five binary factors relating to the previous shout at every time step:

1. Am I still active?
2. Am I buying or selling?
3. Was the last shout accepted or not?
4. Was the last shout a bid or an offer?
5. Would I have traded at the price of the last shout?

This leads to thirty-two distinct scenarios where each scenario is associated with a four element vector $[\beta, \lambda, \rho, \epsilon]$ which defines the price update rule as shown in (6).

$$\Delta_t = [p_t - (q_t \times (1 + \rho) + \epsilon)] \times (1 - \lambda) \times \beta + \lambda \times \Delta_{t-1} \qquad (6)$$

Each element of the vector is set by a uniform random distribution between a maximum and minimum provided at the outset. This means that each simulation has $2^5 \times 4 \times 2 = 256$ parameters associated with price updates. An additional four parameters are associated with the initial profit assumption which leads to this algorithm being named 'ZIP-260', in keeping with the naming convention of the ZIP variants.

One major difference between using ZIP in finance and in electricity markets is the notion of trading at a loss. In finance, it is possible to withold selling a unit to avoid trading at a loss if required. In electricity systems, once a unit of electricity is produced it must be distributed and consumed. Clearly, a supplier could store the energy but currently this has limited large-scale availability. As such, we assume the supplier has no storage capabilities and so must dispose of the electricity generated in some way once it is produced. We also assume that the supplier would sell the electricity at a loss instead of incurring the costs associated with disposing of generated electricity. ZIP shows this behaviour because there is a strict budget constraint on the agents in the form of the limit price, $\lambda_i$, which sets an absolute value on the price they are willing to meet. This represents the maximumim value that a consumer will buy at and a minimum price that suppliers will sell at. Removing this constraint would return the ZIP algorithm to behaviour characterised by the ZI agents [10]. To resolve this issue we allow agents to enter a loss accepting state in certain situations. After updating the price as described earlier (6), a second simple algorithm using a Markov chain controls the state of the agent. If a trade was achieved at time $t$ then all the agents will revert to profit-seeking behaviour. However, if a shout at time $t-1$ was unmet then there is a small probability $(P_L)$ that an agent will return to a loss accepting state. The internal profit assumption remains unchanged in this state but they accept any price that is quoted on the market. Figure 1 shows the distribution of unmatched supply and demand with and without the loss accepting mechanism. Positive values of units before simulation indicate a supplier while a negative value of units indicates a consumer. Figures 1(c) and 1(d) shows how well the resources are allocated. In these simulations, there is an overall deficit in supply.

### 3.3   Price and Fitness Functions

Two general pricing schemes are used throughout the paper; constant and varying demand price. The price for the $i$th agent's $j$th unit is given by $\lambda_{i,j}$. The constant price function is give by (7) whilst the varying price is given by (8) where $u_i$ is a utility function parameter. The suppliers price has been set to be a constant value of 1.5, which, while being simplistic is justified by considering only one trading day where the cost of fossil fuel powered energy and wind powered energy is assumed to be stable.

$$\lambda_{i,j} = 2.5 \tag{7}$$

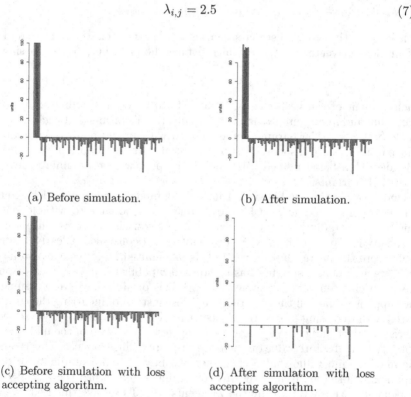

(a) Before simulation.

(b) After simulation.

(c) Before simulation with loss accepting algorithm.

(d) After simulation with loss accepting algorithm.

**Fig. 1.** Distribution of units to agents before and after the simulation runs for a single day, with and without a loss accepting algorithm. Positive units indicate an excess of supply whist negative units indicate an excess of demand.

$$\lambda_{i,j} = 1.20 + \frac{10}{9+j} + u_i(0.05) \tag{8}$$

Throughout the paper two measures of fitness are used, Smith's $\alpha$ [14] and we introduce a measure of unit variance ($\sigma^2$) which we define in (9) where $N$ is the number of consumers and $x$ is a vector representing the units that an agent demands and $\bar{x}$ is the population mean.

$$\sigma^2 = \frac{1}{N} \sum_i^N (x_i - \bar{x})^2 \tag{9}$$

## 4  Simulation Scenarios and Results

We start by defining the supply patterns for Supplier F and Supplier W. In the case of Supplier F we simplified the plant to supply a constant and highly predictable power output, mirroring a fossil fuel plant. On the other hand, Supplier W has been modelled using a Weibull distribution with a probability distribution function as shown in (10). Where $c$, the scale parameter, and $k$, the shape parameter, take values $5.428ms^{-1}$ and $1.4128$ [12] respectively. The power generated, $P(v)$, by a turbine is not a direct mapping from this distribution. A turbine has a cut-in speed ($v_{cut-in}$), below which it will not generate any power; a rated speed ($v_{rated}$), at which it produces maximum power; and a cut-out speed ($v_{cut-out}$), above which it will not produce power (for safety reasons). Between the cut-in and rated wind-speeds, it is assumed here that the power generated is a linear interpolation between zero and the rated power output as shown in (11). For the simulation analysis Supplier W was assumed to own a small wind farm consisting of five turbines. Each turbine is modelled as behaving according to (10) and (11) with no interactions between turbines. MICON turbine specifications [16] as shown in Table 1 were taken to simulate power output for Supplier W sampled half-hourly over a 24 hour period. This gives the wind generation a more unpredictable production pattern as would be the case when integrating renewable energy to the grid.

$$f(v,c,k) = \frac{k}{c}\left(\frac{v}{c}\right)^{k-1} e^{-\left(\frac{v}{c}\right)^k} \qquad v > 0; c > 0; k > 0 \tag{10}$$

$$P(v) = \begin{cases} 0 & 0 \le v < v_{cut-in} \\ P_{rated} \times \frac{v - v_{cut-in}}{v_{rated} - v_{cut-in}} & v_{cut-in} \le v < v_{rated} \\ P_{rated} & v_{rated} \le v < c_{cut-out} \\ 0 & v \ge v_{cut-out} \end{cases} \tag{11}$$

### 4.1  ZIP Performance

We implement the ZIP60 algorithm using a fixed price mechanism with Supplier F as the sole source of electricity. Figure 2(a) shows the demand and supply positions before and after trading for the entire day whilst Figure 2(b) shows the

**Table 1.** MICON TURBINE PARAMETERS

| Parameter | Value |
|---|---|
| $v_{cut-in}$ | $4ms^{-1}$ |
| $v_{rated}$ | $14ms^{-1}$ |
| $v_{cut-out}$ | $25ms^{-1}$ |
| $P_{rated}$ | $200kW$ |

corresponding average price that trades were executed at. This indicates there are two key issues with the simulation, firstly, the fixed price function is very simplistic and allows monopolistic tendencies. Secondly, it indicates a problem with the $\alpha$ measure itself in very uneven market conditions; as long as there is a consistent excess of demand over supply the equilibrium price will favour the supplier. While not shown here, a simulation for only Supplier W showed the same behaviour.

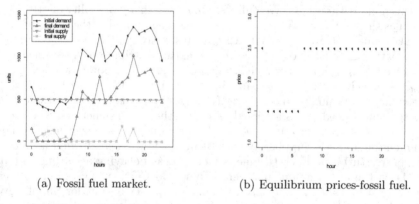

(a) Fossil fuel market.          (b) Equilibrium prices-fossil fuel.

**Fig. 2.** Units traded and average trading price over one day

## 4.2   ZIP-260

We first implement the ZIP-260 algorithm with a fixed price function and without the loss-accepting behaviour. Figures 3(a) and 3(b) show that ZIP-260 behaves similarly to ZIP60 when optimised for $\alpha$ which means that by changing the fitness function we have not adversely affected the market mechanism. Figure 4 shows ZIP60's behaviour when optmised for $\alpha$ and $\sigma^2$ under a variable pricing function which achieves worse $\alpha$ performance than the ZIP-260 algorithm shown in Figure 5.    Having shown that the ZIP-260 algorithm is in fact an improvement on the ZIP60 variant, and that the fitness function $\sigma^2$ is preferred to $\alpha$ as it encourages all available units to be traded, we implement the ZIP-260 algorithm as outlined in §3.2 and introduce $\sigma^2$ as the fitness parameter as well as a variable pricing function. We examine a system where both suppliers are operating to investigate whether fossil fuels can complement wind in this algorithmic market. Figure 6 shows the supply and demand positions before and after trade

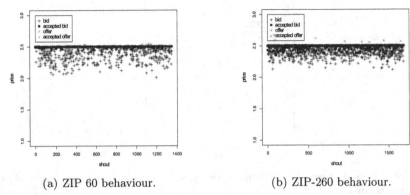

(a) ZIP 60 behaviour.              (b) ZIP-260 behaviour.

**Fig. 3.** The trading activity for hour 12 of the fossil fuel market for both ZIP60 and ZIP-260 algorithms

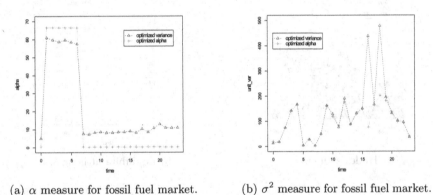

(a) $\alpha$ measure for fossil fuel market.      (b) $\sigma^2$ measure for fossil fuel market.

**Fig. 4.** Performance of the ZIP60 fossil fuel market for the two fitness measures under variable pricing

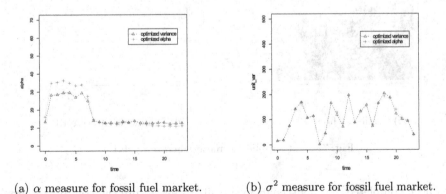

(a) $\alpha$ measure for fossil fuel market.      (b) $\sigma^2$ measure for fossil fuel market.

**Fig. 5.** Performance of the ZIP-260 fossil fuel market for the two fitness measures under variable pricing

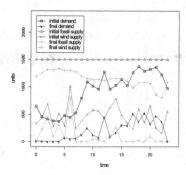

**Fig. 6.** Supply and demand positions of the joint market shown before and after trading has occured

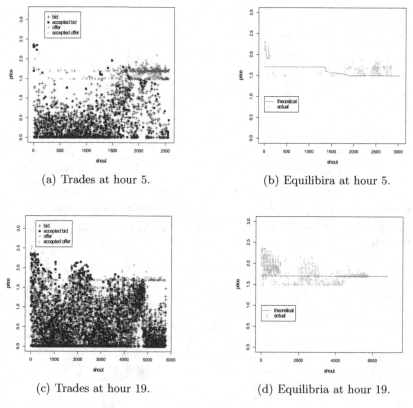

(a) Trades at hour 5.

(b) Equilibira at hour 5.

(c) Trades at hour 19.

(d) Equilibria at hour 19.

**Fig. 7.** Joint supplier scenario where $P_L = 0.1$

has been completed. In this scenario we deliberately ensure that there is an excess of supply in the market in order to test the algorithm's ability to distribute all units of supply. Figure 7 shows the trading activity and price equilibria for different trading periods of the system when the probability of accepting a loss ($P_L$) is set at 0.1. The consumers are constantly shouting zero priced bids which the loss-accepting suppliers are matching. This shows that the consumers are using their collective bargaining power to set the price very low because they have learned that the suppliers will accept it. We then reduce the probability of accepting a loss ($P_L = 0.001$) and observe the trades and equilibria at the same periods, shown in Figure 8. This now shows a more equitable market with the equilibria following the price curve of the consumers.

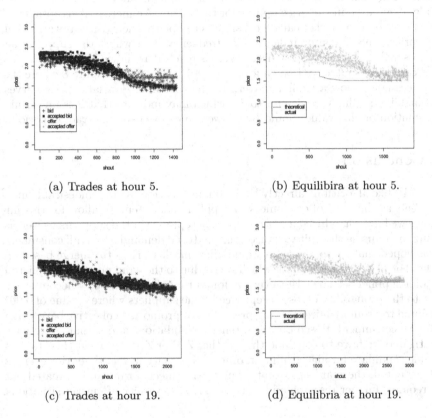

(a) Trades at hour 5.  (b) Equilibira at hour 5.

(c) Trades at hour 19.  (d) Equilibria at hour 19.

**Fig. 8.** Joint supplier scenario where $P_L = 0.001$

## 5   Analysis

We show that the direct application of automated trading in the traditional sense of profit maximizing is actually not appropriate for electricity markets as it leads to monopolistic behaviour by the suppliers. The imbalance of consumers and suppliers seriously affects the performance of the algorithm as the suppliers display monopolistic tendencies. To overcome this issue, we introduced a notion of unit variance as a fitness function which optimises for an equal distribution of outstanding demand units across all the consumers which allows for all available supply units to be allocated, as shown by Figure 1. We compare the performance of the new ZIP-260 algorithm with ZIP60 and see that both measures of performance ($\alpha$ and $\sigma^2$), are greatly improved which can be observed by comparing Figures 4 and 5. Unfortunately, when this is implemented in a joint supplier scenario with the probability to accept a loss set at 0.1 ($P_L = 0.1$), we see the consumers learning to abuse this loss making behaviour by consistantly shouting zero priced bids. This indicates that the collective bargaining power can have a large effect such that the suppliers are not being compensated at all for the electricity they produce. This can be explained by considering the cumulative effect of the learning algorithm used in the price update for the agents. An increased acceptance of lower prices on the market causes consumers to update their prices towards that lower price. This is compounded by the frequency with which the suppliers accept loss making prices, until a price of zero is reached as shown by Figure 7. This effect can be alleviated by reducing $P_L$ to 0.001, as shown in Figure 8 where the bid prices now follow a similar pattern to the consumers' variable price curves. This highlights a high sensitivity to $P_L$ which may indicate that it is an unsuitable solution or this value should be allowed to evolve over the trading period.

## 6    Conclusion

Using ZIP like algorithms directly leads to non-functioning and undesirable market effects as the ratio of consumers to suppliers is uneven. To allow for this imbalance we have introduced a notion of loss-acceptance in the suppliers as well as aiming to minimise the unit variance of un-matched demand across all consumers. It was hoped that these additions would allow market prices to emerge based on the level of supply in the market which varies due to the wind powered supplier in the joint supplier scenario. However, we found that the market now becomes sensitive to the probability of loss acceptance by the suppliers where a value of 0.001 was found to promote fairer trading prices by not promoting collective bargaining power by consumers. Based on the extensive simulations and scenarios we have investigated we have to conclude that neither ZIP or ZIP-260 is, at this stage, a suitable candidate for algorithmic trading of electricity in real time. However, as we have noted, there are signs that a fair market mechanism can be created but this requires further research and testing on a wide variety of market situations.

## References

1. Cliff, D.: The Flash Crash of May 6 2010: WTF? Technical report, Department of Computer Science, University of Bristol, Bristol (2010)
2. Nuti, G., Mirghaemi, M., Treleaven, P., Yingsaeree, C.: Algorithmic Trading. Computer 44, 61–69 (2011)
3. Amin, S.M., Wollenberg, B.F.: Toward a Smart Grid. IEEE Power & Energy Magazine 3(5), 34–41 (2005)
4. Albadi, M.H., El-Saadany, E.F.: Demand Response in Electricity Markets: An Overview. In: 2007 IEEE Power Engineering Society General Meeting, pp. 1–5 (June 2007)

5. Cliff, D.: Minimal-Intelligence Agents for Bargaining Behaviors in Market-Based Environments. Technical report, Hewlett-Packard Laboratories, Bristol (1997)
6. Ramchurn, S.D., Vytelingum, P., Rogers, A., Jennings, N.: Agent-Based Control for Decentralised Demand Side Management in the Smart Grid. In: International Conference on Autonomous Agents and Multiagent Systems AAMAS, Taiwan, pp. 2–6 (2011)
7. Mohsenian-Rad, A.H., Wong, V.W.S., Member, S., Jatskevich, J., Schober, R., Leon-garcia, A.: Autonomous Demand-Side Management Based on Game-Theoretic Energy Consumption Scheduling for the Future Smart Grid. System 1(3), 320–331 (2010)
8. Vytelingum, P., Ramchurn, S.D., Voice, T.D., Rogers, A., Jennings, N.R.: Trading Agents for the Smart Electricity Grid. In: van der Hoek, Kaminka, Lesperance, Luck, Sen (eds.) 9th Int. Conf. on Autonomous Agents and Multiagent Systems (AAMAS 2010), Toronto, pp. 897–904 (2010)
9. Tesfatsion, L., Judd, K.L. (eds.): Handbook of Computational Economics. North-Holland
10. Gode, D.K., Sunder, S.: Allocative Efficiency of Markets with Zero-Intelligence Traders: Market as a Partial Substitute for Individual Rationality. The Journal of Political Economy 101(1), 119–137 (1993)
11. Elexon: The Electricity Trading Arrangements: A Beginner's Guide. Technical Report July, Elexon (2009)
12. Qian, K., Zhou, C., Li, Z., Yuan, Y.: Benefits of energy storage in power systems with high level of intermittent generation. In: 20th International Conference on Electricity Distribution. Number 0358, Prague, pp. 8–11 (2009)
13. CER: Electricity Smart Metering Customer Behaviour Trials (CBT) Findings Report. Technical report, The Commission for Energy Regulation (2011)
14. Smith, V.L.: An Experimental Study of Competitive Market Behavior. Journal of Political Economy 70(2), 111–137 (1962)
15. Cliff, D.: Zip60: An enhanced variant of the zip trading algorithm. In: E-Commerce Technology. In: The 3rd IEEE International Conference on the 8th IEEE International Conference on and Enterprise Computing, E-Commerce, and E-Services (June 2006)
16. Yeh, T.H., Wang, L.: A Study on Generator Capacity for Wind Turbines Under Various Tower Heights and Rated Wind Speeds Using Weibull Distribution. IEEE Transactions on Energy Conversion 23(2), 592–602 (2008)

# A Semantic-Based Algorithm for Data Dissemination in Opportunistic Networks*

Marco Conti, Matteo Mordacchini,
Andrea Passarella, and Liudmila Rozanova

IIT – CNR, Pisa, Italy
{firstname.lastname}@iit.cnr.it

**Abstract.** The opportunistic data dissemination problem for mobile devices is an open topic that has attracted many investigations so far. At the best of our knowledge, none of these approaches takes into account the semantic side of the data shared in an opportunistic network. In this paper, we present an algorithm that, starting from the semantic data annotations given by the users themselves, builds a semantic network representation of the information. Exploiting this description, we detail how two different semantic networks can interact upon contact, in order to spread and receive useful information. In order to provide a performance evaluation of such a solution, we show a preliminary set of results obtained in a simulated scenario.

## 1 Introduction

The increasing, pervasive presence of devices interacting among themselves and their users is leading to a complex and vast information environment, where information flows from the physical world to the cyber one, and vice-versa. Users mobile devices, sensor networks, and all the devices spread in the environment with data generation capabilities (e.g., in Internet of Things applications) will constantly generate huge amounts of data thus generating a very rich information landscape. This scenario is known as the *Cyber–Physical World* (CPW) convergence [1]. Mobile devices will act in the CPW convergence scenario as *proxies* of their human users. They will be in charge of discovering and evaluating the relevance for their human users of the huge amount of information available in the cyber world. This has to be done quickly and using limited resources, as devices will be constantly face large amounts of data items. This situation resemble what the *human brain* does when it has to assess the relevance of the information coming from the surrounding environment. Since devices will act as the *avatars* of their owners, considering the way the human brain deals with huge amounts

---

* This work is funded by the EC under the RECOGNITION (FP7-IST 257756) and EINS (FP7-FIRE 288021) projects. This work was carried out during the tenure of L. Rozanova's ERCIM "Alain Bensoussan" Fellowship Programme. This Programme is supported by the Marie Curie Co-funding of Regional, National and International Programmes (COFUND) of the European Commission.

W. Elmenreich, F. Dressler, and V. Loreto (Eds.): IWSOS 2013, LNCS 8221, pp. 14–26, 2014.

of data in a short time is a sensible point for designing effective and efficient information dissemination schemes in the CPW convergence scenario .

Opportunistic networking [2] is one of the key paradigms to support direct communication between devices in scenarios like the CPW convergence. In this paradigm, nodes are mobile, and forwarding of messages occurs based on the store, carry and forward concept. In this paper we present a data dissemination algorithm for opportunistic networks inspired by real human communication schemes. It exploits the *semantic* representation of data (e.g., tags and other metadata associated to data items) in order to assess the relevance of information to be exchanged among mobile nodes upon contact. A dissemination scheme for spreading the actual content associated with semantic data can be built on top of such an algorithm, as we point out later in Sec. 3. The focus of this paper is to define and give an initial performance evaluation of a semantic data dissemination mechanism, in order to study the viability of this approach before exploiting it to design effective and efficient content dissemination schemes in opportunistic networks. In this proposal, the data stored in a device is semantically represented using an approach based on semantic networks. A semantic network is a graph, where vertices are the semantic concepts (e.g., the tags associated to data items) and edges represent the connections that exist among them (e.g., the logical link between two tags). We describe how semantic networks can be used to determine the relevance of information to be exchanged upon physical contact among nodes. Essentially, similar to a real human communication, the selection of information starts from a set of semantic concepts in common between the two nodes. The relevance for one node of the rest of the information available in the other node's semantic network can then be assessed based on the links between the semantic concepts in the semantic network. Similarly to a human dialogue, information logically closer to a set of common concepts is exchanged first, and the longer the discussion (i.e. the duration of the physical contact among nodes), the greater the amount of information exchanged. Once an encounter finishes, new semantic concepts are passed from one node to the another, and new connections among new and old concepts can be established, thus increasing the knowledge of each node.

The rest of this paper is organized as follows. In Sec. 2 we review the relevant literature for the opportunistic data dissemination problem. In Sec. 3 we give a general overview of the concepts behind the solution we propose, while in Sec. 3.1, Sec. 3.2 and Sec. 3.3 we describe how semantic networks can be constructed and interact. Sec. 4 presents some preliminary simulation results obtained with this solution. Finally, Sec. 5 concludes the paper.

## 2 Related Work

The data dissemination problem in opportunistic networks has been faced by many solutions in literature. PodNet [3] is one of the first works on this subject. The authors of PodNet propose four different strategies to weight the relevance of data to be exchanged on the basis of the estimated popularity of the general topic (*channel*) the data belongs to. More refined approaches try to take advantage of the information about users' social relationships to drive the dissemination process. For instance, in [4], the authors propose to build a pub/sub overlay in an opportunistic network. The most "socially-connected" nodes, i.e., those nodes that are expected to be most available and easily reachable in the network, take the role of brokers, as in more traditional pub/sub systems. They are in charge to collect information from their social groups, spread this data among them, and eventually deliver it toward interested peers. Also the authors of ContentPlace [5] propose to drive the data dissemination process using the social structure of the network of users. Specifically, each node tries to fetch from other encountered peers those data items that are likely of interest to other users it has social relationships with (and which, therefore, are expected to be in touch with them in the near future). In [6,7], the authors define a data dissemination scheme that directly embodies and exploits the very same cognitive rules (*cognitive heuristics*) used by the human brain to assert the relevance of information. This solution proves to be as effective as another scheme like ContentPlace in disseminating the information, while requiring much less overhead.

With respect to all these approaches, in this paper we take a different direction. None of the approaches above take into consideration the semantic dimension of data. In the following we define how a semantic representation of information can be used to determine the relevance of information to be exchange in an opportunistic network scenario, and we validate this proposal with preliminary simulation results.

## 3   Data Dissemination Using Semantic Networks

In order to semantically represent the information carried by each user, we exploit an approach based on semantic networks. A semantic network is a graph where vertices are semantic concepts and edges connect semantically related concepts. In order to derive this representation, we consider that each user has a collection of data elements, like pictures, blog posts, status updates, geospatial data, etc. Some of these data items could be difficult to analyze semantically. Anyway, we could take advantage of the fact that many of these data items are usually associated with tags, as happens in many real-world social networks like Twitter, Flickr, Instagram, etc. Thus, a simple and, at the same time, effective way of creating a semantic network is to use data tagging, as depicted in Fig. 1. Tags can then become the nodes of a user semantic network, while edges can be derived from the associations the user has done when tagging their own data.

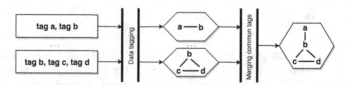

**Fig. 1.** Creation of a Semantic Network by data tagging

Note that not all the information belonging to a user may by represented in the user's semantic network. It seems reasonable that only information that is assumed to be relevant for the user at a given time is mapped in her semantic network. In order to select this information, we can act in a similar way like the human brain does, exploting an approach built on models of the human forgetting process. Thus, we can assume that information that has not been used for a long time is forgotten, while information that is frequently accessed is harder to forget. Rather than a limit, forgetting is a mechanism that could aid some human cognitive processes (e.g. [8]) by letting them keep and consider only the most relevant data. To model a forgetting process into our system, we propose to introduce a special "forgetting" function similar to an experimental curve that has been first obtained by H. Ebbinghaus in 1885 [9] and then confirmed and refined by numerous subsequent studies (Fig. 2). This curve reflects the exponential decline of individual human memory retention with time. The rate of forgetting depends on the repetition rate of incoming information.

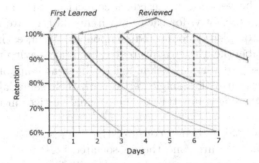

**Fig. 2.** Example of a forgetting curve

Given the description above, the semantic network of each user is dynamic; it is constructed from the individual information by tagging, taking into account the relationships between tags defined by the user herself, and frequency and/or duration of accessing them.

Using this description, in the following sections we first describe how each device can build its own semantic network from the data it owns. Next, we exploit the semantic network associated to data items in order to define a data dissemination scheme in an opportunistic network scenario, where information is exchanged only upon physical contact between nodes. The algorithm we describe deals with semantic data only. Anyway, it can be easily used as the basis for designing the dissemination of the actual content the semantic data is related to, as we stated in the Introduction. The semantic data exchanged with another node can be used to sort the content items in order to select which of them can be transferred to the encountered peer. For istance, a simple way to do this can be to give precedence to those items with the largest intersection between the semantic concepts associated to them and the set of semantic concepts that are going to be exchanged with the other peer. In this paper, we describe how semantic data can be spread in an opportunistc network scenario, leaving the design of an effective and efficient content dissemination mechanism based on this scheme as a future research direction.

## 3.1  Semantic Network Creation

We define the semantic network of each user as a *dynamic weighted graph* $G = \{V, E, f(e,t)\} : t \in T$, where $t$ is the time, $V$ is the set of vertices in the graph, $E$ is the set of edges and $f(e,t)$ is a weighting function for each $e \in E$ that reproduces the human forgetting process in our system. In addition, each edge $e_{ij}$ between two vertices $i$ and $j$ has an associated "popularity" $p_{ij}^t$ that measures the number of times $e_{ij}$ was present in exchanges with other nodes' semantic networks in the encounters happened till time $t$.

In the context of an active user participation to the creation of content, we assume that each data item owned by a user is associated with a set of tags, defined by the user herself. In the graph $G$ that represents the user's semantic network, tags are the vertices, while edges connecting tags are created using the following strategy. Firstly, for each data item, its tags are linked together in order to form a completely connected component. Then, each set of vertices carrying the same label (i.e. they where created from tags having the same name) is considered. These are vertices belonging to different components and they are merged together, forming a single vertex with their same label. This single vertex inherits all the edges pointing to the original vertices in their respective components.

As an example of this process, consider the example given in Fig. 3. The user has two different pictures and their associated tags. For the two pictures, two completely connected components are created. Then, the components are merged using the common vertex "lake" as the pivot of this process.

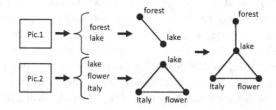

**Fig. 3.** Creation process of a user Semantic Network

## 3.2  Forgetting Mechanism

Let us now define the forgetting function $f(e,t)$ as a function that is able to assign a weight to any edge $e$ at a time $t$. If $f(e,t) \leq f_{min}$, where $f_{min}$ is a limiting threshold value, then the edge $e$ does not exist in the semantic network at time $t$. Initially, at time $t_0 = 0$, for any $e \in E$, we have that: $f(e,t_0) = 1$. Subsequently, for any edge $e_{ij}$ and time $t > t_0$, in each interval $(t^*, t)$, where

$t^*$ is the last time this edge was used in exchanges with other peers (i.e. its last "activation"):

$$f(e_{ij}, t) = \alpha e^{-\beta_{ij}(t-t^*)}$$

where $\alpha$ is a normalizing coefficient and $\beta_{ij}$ is the "speed of forgetting", i.e. the weakening of the connection (taken in accordance with the experimental curve obtained by H. Ebbinghaus [9]). Obviously, $\beta_{ij}$ depends on the total number of previous connections. Then the "popular" connections are "being forgotten" more slowly in the situation when there are no subsequent connections. So, we can define this parameter as follows: $\beta_{ij} = \frac{\beta}{p_{ij}^t}$ , where $\beta$ is a speed coefficient and $p_{ij}^t$ is the "popularity" of $e_{ij}$ at time $t$, i.e. the number of times $e_{ij}$ has been used in the encounters happened before $t$. Fig. 4 shows one of the side effects of the forgetting process. Not only edges, but even vertices can be forgotten. In fact, if the deletion of an edge $e$ implies that a vertex $v$ at one of $e$'s endpoints is no longer connected to any other vertex in the network, then $v$ is also deleted from the semantic network. After that, in order for it to reappear in the semantic network, it should be received during successive encounters with other nodes, using the data exchange scheme detailed in the following section.

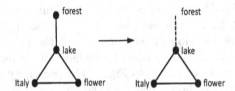

**Fig. 4.** Effects of the forgetting process

## 3.3   Interaction between Semantic Networks

In a real human communication, a dialogue begins with some concepts that are in common between both parties. Similarly to this behaviour, we let the interaction between two semantic networks start with one or few key concepts (vertices) that belong to both semantic networks. Starting from these key vertices, each device is able to compute which are the vertices and edges from its own semantic network that are to be communicated and transferred to the other party. Precisely, looking from the viewpoint of a device, let its semantic network $G = (V, E, f(e, t))$ be the *donor network*, while the other party semantic network $G' = (V', E', f'(e', t))$ is termed the *recipient network*. During the interaction between these two networks, concepts of the donor network are included in the recipient network and new connections between concepts are formed. As a result we obtain a *new (updated) recipient* semantic network $G^* = (V^*, E^*, f^*(e^*, t))$. Being $|V| = n$ and $|V'| = m$, we have that $|V^*| = l \leq n + m$.

In order to determine which vertices and edges from the donor network are exchanged during a communication, a node computes a *contributed network* $C = (\bar{V}, \bar{E}, \bar{f}(\bar{e}, t))$, which will contain the data that will be transmitted. Once the contributed network is received, it is merged with the recipient network in order to create the updated recipient network. In the following, we describe how a node computes the contributed network and how this is finally merged with the recipient network. Hereafter, when we say that an edge is included in the contributed network, we imply that the vertices at both the endpoints of that edge are also included, in case the contributed network does not already contain them.

Supposing that the interaction starts at time $t$, the pseudo-code used for initializing the contributed network is presented in Alg.1. As already stated, we assume that a data exchange between two nodes starts from a set of shared semantic concepts, i.e. a set of vertices $K = \{v_k : v_k \in V \cap V'\}$ (lines 1–6 of the pseudo-code). Note that in case $K = \emptyset$, nothing is passed from one node to the other. Vertices and the corresponding edges that connect them directly (one-hop distance) with key nodes are included instantly in the contributed network (lines 7–12). This process is analogue to the idea of *"gestalt"*, when the understanding is not limited to a single concept, but brings a system of linked concepts i.e. a part of the semantic network. Note that the weights of edges included in the contributed network are set to the maximum in both the donor and contributed networks (lines 10–11), as the exchange of this data leads to a sort of "activation" of the corresponding edge in memory, thus inducing the forgetting process to "restart". In order to compute which of the remaining vertices and edges of the donor network should be included in the contributed network, we proceed as detailed in Alg. 2. Edges will be subject to a "warming" process, that will mainly depend on the duration of the contact and the proximity of an edge's endpoints to a key vertex. Thus, vertices and edges will be included in the contributed network by levels, as shown in Fig. 5. The proximity value is computed as the minimum number of hops needed to the reach any of the key vertices in the donor network. Edges that connect vertices that are closer to a key node will be "warmed up" faster than those linking vertices located far away. Moreover, the longer will last an interaction between two nodes, the easier an edge will be "warmed". This process mimics what happens in a human communication process, where, starting from a set of common concepts, other semantically connected notions could be included in the dialogue. The longer the discussion

---

**Algorithm 1.** Contributed Network Creation at time $t$

```
1: Let G = (V, E, f(e, t)) be the donor network;
2: Let C = (V̄, Ē, f̄(ē, t)) be the contributed network;
3: Let K be the set of key vertices, K ⊆ V
4: for each k ∈ K do
5:     V̄∪ = k
6: end for
7: for each v ∈ V and each k ∈ K such that ∃e_{kv} ∈ E do
8:     V̄∪ = v
9:     Ē∪ = e_{kv}
10:    Set f(e_{kv}, t) = 1 in G
11:    Set f̄(e_{kv}, t) = 1 in C
12: end for
```

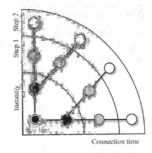

**Fig. 5.** Selection of vertices from the donor network using proximity levels

takes, the more concepts are exchanged. When the interaction is terminated, the contributed network has only those edges (and related vertices) that exceed a "warm" activation threshold. Thus, only that information is transferred to the recipient network.

In detail, supposing that a connection starts at time $t$ and ends at time $t^*$, edge warming is computed using the following formula:

$$w(e_{ij}, \Delta_t) = \frac{\gamma_{step}}{1 + e^{-p_{ij}^{t^*}}}(1 - e^{-\tau \Delta_t})$$

where $e_{ij} \in E$ is an edge of the donor network, $\Delta_t$ is the duration of the connection, i.e. $\Delta_t = t^* - t$, $p_{ij}^{t^*}$ is the popularity of $e_{ij}$ at time $t^*$ and $\tau$ is a normalizing factor. The coefficient $\gamma_{step}$ is used to weight the proximity of $e_{ij}$ to any key vertex. We can define this value as $\gamma_{step} = \frac{\gamma}{n}$, where $n$ is the the number of hops in the shortest path to the nearest key vertex and $\gamma$ is a normalizing factor. Being $w_{min}$ the minimum warm threshold, the contributed network will contain an edge $e_{ij}$ **iff** $w(e_{ij}, \Delta_t) \geq w_{min}$ (lines 10–17). Moreover, in order to limit the amount of information exchanged during an encounter, we consider that, apart from its warm weight, an edge $e_{ij}$ is included in the contributed network **iff** it is within $h$ hops from any key vertex in the donor network (line 8). At the end of the interaction, the contributed network is transferred from the donor node to the recipient one. The contributed network is merged with the recipient network using Alg. 3. Edges and vertices that do not exist in the recipient network are added (lines 5–14). In case an edge of the contributed network already exists in the recipient one (line 16), its weight is set to the maximum between the weigth already assigned to it in the recipient network and the weight passed by the contributed network.

## 4    Simulation Results

In order to evaluate the proposed algorithm, we simulated its behaviour in the following scenario. We consider 99 mobile nodes that move in a 1000mx1000m area. In order to simulate real user movement patterns, nodes move according to the HCMM model [10]. This is a mobility model that integrates temporal,

**Algorithm 2.** Contributed Network computation at the end of an encounter

1: Let $G = (V, E, f(e, t))$ be the donor network;
2: Let $C = (\bar{V}, \bar{E}, \bar{f}(\bar{e}, t))$ be the already initialized contributed network;
3: Let $K$ be the set of *key vertices*, $K \subseteq V$
4: Let $A = (V \cap \bar{V}) - K$
5: Let $h$ be the depth limit
6: Let $w_{min}$ be the weight threshold
7: Let $depth = 2$
8: **while** $depth \leq h$ **do**
9:    Let $B = \emptyset$
10:    **for** each $v \in (V - \bar{V})$ such that $\exists e_{av} \in E, a \in A$ **do**
11:       **if** $w(e_{av}, t^*) \geq w_{min}$ **then**
12:          $\bar{V} \cup = v$
13:          $\bar{E} \cup = e_{a,v}$
14:          Set $f(e_{av}, t) = 1$ in $G$
15:          Set $\bar{f}(e_{av}, t) = w(e_{av}, \Delta_t)$ in $C$
16:          $B \cup = v$
17:       **end if**
18:    **end for**
19:    $A = B$
20:    $depth = depth + 1$
21: **end while**

**Algorithm 3.** Merging of contributed and recipient networks

1: Let $G' = (V', E', f'(e', t))$ be the recipient network;
2: Let $C = (\bar{V}, \bar{E}, \bar{f}(\bar{e}, t))$ be the contributed network;
3: Let $G^* = (V^*, E^*, f^*(e^*, t))$ be the updated recipient network;
4: $G^* = G'$;
5: **for** each $\bar{e}_{ij} \in \bar{E}$ **do**
6:    **if** $\bar{e}_{ij} \notin E^*$ **then**
7:       **if** $\bar{v}_i \notin V^*$ **then**
8:          $V^* \cup = \bar{v}_i$
9:       **end if**
10:       **if** $\bar{v}_j \notin V^*$ **then**
11:          $V^* \cup = \bar{v}_j$
12:       **end if**
13:       $E^* \cup = \bar{e}_{ij}$
14:       Set $f^*(\bar{e}_{ij}, t) = \bar{f}(\bar{e}_{ij}, t)$
15:    **else**
16:       Set $f^*(\bar{e}_{ij}, t) = \max(f^*(\bar{e}_{ij}, t), \bar{f}(\bar{e}_{ij}, t))$;
17:    **end if**
18: **end for**

social and spatial notions in order to obtain an accurate representation of real user movements. Specifically, in HCMM the simulation space is divided in cells representing different social communities. Nodes move between social communities, and nodes movements are driven by social links between them. In this preliminary study, we consider that there exists only one social community, i.e.

the simulation space consists of one cell that covers all the simulation area. Data assigned to these nodes is selected from the CoPhIR dataset [11]. This is a dataset containing more than 100 million images taken from Flickr. Along with other data, for each image it is possible to know the user that generated it and the associated tags. In order to create a useful dataset to test our solution, we selected images with at least 5 tags each. This number was chosen considering that the overall mean number of tags per image in the dataset is 5.02. Then, we extracted those users that have at least 10 such images in their collections. Finally, from this set of users, we randomly chose the 99 users that we used in the simulation. For each of these users, a corresponding semantic network is created, according to the description given in Sec. 3.1. We then study the transient state of the interaction between these users, by repeating 10 different tests obtained by producing 10 different mobility traces using the HCMM model. Results reported in the following simulations are the average of all the performed tests. Each simulation experiment runs for 5000 sec.

Fig. 6(a) (log-scale on the $x$ axis) shows the evolution over time of the tags Hit Ratio for three different settings of the forget function. The overall Hit Ratio is defined as the mean of the per tag hit ratio. This latter quantity is computed as the mean number of nodes having a given tag in their semantic networks at a given time. We set the parameters of the forget function in order to have the less popular edges be deleted after 50, 100 and 250 sec of inactivity, respectively. In this scenario, vertices at more than 3 hops from a key vertex are not exchanged during contacts. The first thing to note is that tags are not spread to any node in all the cases we present. This is consistent with the view that each user is probably not interested in every semantic concept available. Rather, she is more interested in fetching the concepts more pertinent with her own knowledge. Anyway, as one could expect, the forget function plays a key role in the data dissemination process. The longer we let the edges (and related vertices) stay in their semantic networks before being forgotten, the higher will be the final diffusion of the corresponding tags in the network. Since there are great variations in the Hit Ratio values, a proper tuning of the forget function parameters implies

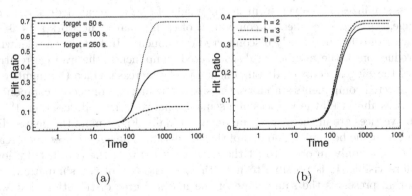

(a)                              (b)

**Fig. 6.** Hit Ratios for different settings of (a) the forget function and (b) parameter $h$

a trade-off between the willingness to let the concepts permeate the network and the need to limit the associated resource consumption (storage, bandwidth, etc.). In all the reported results, the Hit Ratio reaches a stabilization value. This is due to the fact that (i) useful information has been already shared among potentially interested nodes; (ii) the meeting rate of this data is faster than the forget process, i.e. data is seen again before the associated forget function value falls below the forgetting threshold. In Fig. 6(b) we report the impact on the Hit Ratio of the proximity level limit $h$ for the exchange of vertices and related edges. In this case we fixed the forget function parameters in order to let the least popular edges disappear after 100 sec. of inactivity. Results are obtained for $h = 2, 3$ and 5. We can see that allowing to include in the contributed networks a larger (i.e. more distant from key vertices) portion of the donor networks result in a larger diffusion of semantic concepts (i.e. an higher Hit Ratio). Anyway, the Hit Ratio is less sensitive to changing in the $h$ value rather than to changes in the forget function parameters. Thus, although different values of $h$ lead to different Hit Ratios, tuning of this parameter is less critical than that of the forget function, since it leads to relatively small differences in the final Hit Ratio values.

In the next sets of results, we study some of the general properties of the semantic networks as they result at the end of the simulation. Main parameters are: $h = 3$; the forget function deletes the least popular edges after 100 sec. The left side of Fig. 7 shows the evolution over time of the mean number of different connected components that form each semantic network. Each semantic network is not a complete graph. Rather it is a set of different weakly connected components. On one hand, these different sets of semantic concepts (i.e. vertices) may be thought to represent different, semantically uncorrelated groups of topics the user is interested in. Anyway, these sets could also be disconnected one from the others since the user lacks the knowledge needed to put them together. We can see that, as time passes, the mean number of different connected components rapidly falls down, as an effect of the data dissemination process. Since new vertices and new connections between old and new vertices are created, some sets of previously disconnected nodes start to merge, until the process stabilizes. Moreover, from the right side of Fig. 7, we can deduce that, on average, the biggest connected component of each semantic network acts as an attractor of other previously disconnected components. In fact, in parallel with the reduction of the number of disconnected components, the average relative size of the biggest connected component rapidly increase. Once the number of disconnected components stabilizes, the size of the biggest connected component comprises almost all the vertices of a semantic network. Indeed, less then 1% of all the vertices are in the other components. Finally, Fig. 8 plots the degree distribution at the beginning and end of the simulation. The figures uses a log-scale on the $y$ axis only, in order to let the differences between the two distributions be more visible. It is possible to note that, at the end of the simulation, the algorithm preserves the same slope of the nodes degree distribution that was present before the simulation starts. Anyway, there is an increased probability

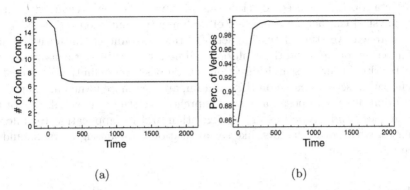

**Fig. 7.** Evolution of (a) the number of distinct connected components and (b) the size of the biggest connected component

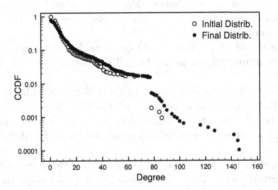

**Fig. 8.** Nodes' degree distribution before and at the end of the simulation

to find nodes with high degree, as shown by the CCDF of the final degree distribution. Node with an intial higher degree have more chances to be involved in an exchange than nodes with lower degrees. Moreover, the forget process cuts less popular edges, thus further reducing the degree of less spread vertices. Edges attached to high-degree nodes take advantage of the nodes' probability to be exchanged in order to avoid to be forgotten. Eventually, this mechanism favours the increase of the degree of already well-connected nodes.

## 5   Conclusions

The semantic information associated with data items could be a powerful tool in a data dissemination scheme in order to assert the relevance and relationship of already owned information and newly discovered knowledge. Exploiting the data semantic tagging done by the users themselves, we defined a semantic-based data dissemination algorithm for opportunistic networks. We show how each device is able to give a semantic network representation of its own data and how this representation can be used to select the information to be exchanged by users upon physical contact. In a first set of preliminary simulation results based on this approach, we studied the impact of various parameters on both the data dissemination process and the evolution and final properties of the users' semantic networks. Future research directions encompass the definition of a content dissemination scheme based on this solution, an even more formal mathematical description of this proposal, a more comprehensive study via simulation of the performances and properties of the algorithm and its application to different scenarios, where other factors, like social relationships among users, should be taken into account.

# References

1. Conti, M., Das, S.K., Bisdikian, C., Kumar, M., Ni, L.M., Passarella, A., Roussos, G., Trster, G., Tsudik, G., Zambonelli, F.: Looking ahead in pervasive computing: Challenges and opportunities in the era of cyberphysical convergence. Pervasive and Mobile Computing 8(1), 2–21 (2012)
2. Pelusi, L., Passarella, A., Conti, M.: Opportunistic networking: data forwarding in disconnected mobile ad hoc networks. IEEE Communications Magazine 44(11), 134–141 (2006)
3. Lenders, V., May, M., Karlsson, G., Wacha, C.: Wireless ad hoc podcasting. SIG-MOBILE Mob. Comput. Commun. Rev. 12, 65–67 (2008)
4. Yoneki, E., Hui, P., Chan, S., Crowcroft, J.: A socio-aware overlay for publish/subscribe communication in delay tolerant networks. In: MSWiM, pp. 225–234 (2007)
5. Boldrini, C., Conti, M., Passarella, A.: Design and performance evaluation of contentplace, a social-aware data dissemination system for opportunistic networks. Comput. Netw. 54, 589–604 (2010)
6. Conti, M., Mordacchini, M., Passarella, A.: Data dissemination in opportunistic networks using cognitive heuristics. In: The Fifth IEEE WoWMoM Workshop on Autonomic and Opportunistic Communications (AOC 2011), pp. 1–6. IEEE (2011)
7. Bruno, R., Conti, M., Mordacchini, M., Passarella, A.: An analytical model for content dissemination in opportunistic networks using cognitive heuristics. In: Proceedings of the 15th ACM International Conference on Modeling, Analysis and Simulation of Wireless and Mobile Systems, MSWiM 2012, pp. 61–68. ACM, New York (2012)
8. Schooler, L., Hertwig, R.: How forgetting aids heuristic inference. Psychological Review 112(3), 610 (2005)
9. Ebbinghaus, H.: Memory: A contribution to experimental psychology, vol. 3. Teachers college, Columbia university (1913)
10. Boldrini, C., Passarella, A.: Hcmm: Modelling spatial and temporal properties of human mobility driven by users' social relationships. Comput. Commun. 33, 1056–1074 (2010)
11. Bolettieri, P., Esuli, A., Falchi, F., Lucchese, C., Perego, R., Piccioli, T., Rabitti, F.: CoPhIR: A test collection for content-based image retrieval. CoRR abs/0905.4627v2 (2009)

# Characteristic Analysis of Response Threshold Model and Its Application for Self-organizing Network Control

Takuya Iwai, Naoki Wakamiya, and Masayuki Murata

Graduate School of Information Science and Technology,
Osaka University, Suita, Osaka 565-0871, Japan
{t-iwai,wakamiya,murata}@ist.osaka-u.ac.jp

**Abstract.** There is an emerging research area to adopt bio-inspired algorithms to self-organize an information network system. Despite strong interests on their benefits, i.e. high robustness, adaptability, and scalability, the behavior of bio-inspired algorithms under non-negligible perturbation such as loss of information and failure of nodes observed in the realistic environment is not well investigated. Because of lack of knowledge, none can clearly identify the range of application of a bio-inspired algorithm to challenging issues of information networks. Therefore, to tackle the problem and accelerate researches in this area, we need to understand characteristics of bio-inspired algorithms from the perspective of network control. In this paper, taking a response threshold model as an example, we discuss the robustness and adaptability of bio-inspired model and its application to network control. Through simulation experiments and mathematical analysis, we show an existence condition of the equilibrium state in the lossy environment. We also clarify the influence of the environmental condition and control parameters on the transient behavior and the recovery time.

**Keywords:** self-organization, response threshold model, robustness, adaptability, linear stability theory.

## 1 Introduction

For information networks to remain one of infrastructures indispensable for a safe, secure, and comfortable society, they must be more robust, adaptive, and scalable against ever-increasing size, complexity, and dynamics. To this end, researchers focus on self-organizing behavior of biological systems, where a global pattern emerges from mutual and local interactions among simple individuals, and develop novel control mechanisms [1].

Bio-inspired control mechanisms not only mimic behavior of biological organisms but are based on nonlinear mathematical models which explain or reproduce biological self-organization. Since bio-inspired mathematical models, which we call bio-models in this paper, are shown to have excellent characteristics, e.g. high convergence and stability, network control mechanisms based on bio-models

W. Elmenreich, F. Dressler, and V. Loreto (Eds.): IWSOS 2013, LNCS 8221, pp. 27–38, 2014.

are expected to be robust, adaptive, and scalable [2–5]. Successful attempts published in literatures support this expectation and there is no doubt about the usefulness of bio-models [6, 7].

However, bio-models are not necessarily versatile. One can achieve the best performance in one environment while it is useless in other. Furthermore, a bio-inspired network control mechanism often experiences a variety of perturbation such as loss of information and failure of node and as a result it would fail in providing intended results in the actual environment. Therefore, we need deep understanding of bio-models especially in regard to their fundamental limits and applicability to network control suffering from perturbations. For example, they evaluated the influence of delay on a bio-inspired synchronization mechanism adopting the pulse-coupled oscillator model and showed that the synchronization error became as much as the propagation delay at the worst cases [8]. In [9], it is shown that the maximum rate of information loss that autonomous pattern formation based on the reaction diffusion model of biological morphogenesis can tolerate was as high as 35%.

In this paper, by taking a response threshold model [10] as an example, we analyze the influence of information loss and node failure on the stability of a response threshold model-based network control. The response threshold model is a mathematical model of division of labors in a colony of social insects. It has been applied to variety of self-organizing network control, such as task allocation [11], topology control [12], routing [13] and cache replacement [14]. First, we evaluate the influence of information loss on autonomous task allocation. Simulation results show that the number of workers increases as the loss rate is higher while the loss rate does not have serious impact on the recovery time from node failure. Next, to clarify the maximum loss rate that response threshold model-based task allocation can tolerate, we conduct mathematical analysis of the model in lossy environment and derive existence conditions of an equilibrium state. Then, we formulate the recovery time and clarify important parameters to shorten it.

The remainder of this paper is organized as follows. First, in section 2 we briefly introduce a response threshold model. Next, in section 3, we evaluate the influence of information loss on transient behavior during recovery from node failures through simulation experiments. Then, in section 4 we build an analytical model and investigate the relationship between parameters and transient behavior. Finally, in section 5, we conclude this paper.

## 2    Mathematical Model of Division of Labors

A response threshold model [10] is a mathematical model which imitates a mechanism of adaptive division of labors in a colony of social insects. A colony is divided into two groups of workers and non-workers based on autonomous decision of individuals. The size of each group is well adjusted to meet the task-associated demand or stimulus intensity. In the following, we consider there is one task to be performed in the colony for the sake of simplicity of explanation.

Let $s(k)$ $(\geq 0)$ be the task-associated stimulus intensity at discrete time step $k$. The stimulus intensity gradually increases over time, and it decreases as individuals work. Its dynamics is formulated by the following discrete equation.

$$s(k+1) = s(k) + \delta - \frac{w(k)}{M} \tag{1}$$

Here constant $\delta$ $(0 \leq \delta \leq 1)$ is the increasing rate of the stimulus intensity. $w(k)$ is the number of workers at time $k$. $M$ $(> 0)$ is the total number of individuals which are capable of performing the task. Based on the model, the stimulus intensity becomes stable when the ratio of workers in the colony is equal to $\delta$. The model can easily be extended to consider the absolute number of workers not the ratio by appropriately defining $\delta$.

By being simulated by the stimulus, each individual stochastically decides whether to perform the task. The state of individual $i$ at time $k$ is denoted as $X_i(k) \in \{0, 1\}$, where 0 means it is a non-worker and 1 does a worker. The probability $P(X_i(k) = 0 \to X_i(k+1) = 1)$ that non-worker $i$ becomes a worker and begins performing the task at time $k+1$ is given by the following equation.

$$P(X_i(k) = 0 \to X_i(k+1) = 1) = \frac{s^2(k)}{s^2(k) + \theta_i^2(k)} \tag{2}$$

Here $\theta_i(k)$ $(> 0)$ is a threshold at time $k$, which corresponds to hesitation of individual $i$ in performing the task. Therefore, an individual with a smaller threshold is more likely to become a worker more often than those with a larger threshold.

The probability $P(X_i(k) = 1 \to X_i(k+1) = 0)$ that individual $i$, a worker, quits working at time $k+1$ is given by constant $p$ $(0 \leq p \leq 1)$.

$$P(X_i(k) = 1 \to X_i(k+1) = 0) = p \tag{3}$$

Quitting a task at the constant rate enables rotation of the task among individuals, that is, work-sharing or load balancing. Given $p$, the average duration that an individual performs the task becomes $1/p$.

When the number of idle individuals occasionally increases by addition of newcomers or the number of workers decreases for sudden death, the stimulus intensity eventually increases and individuals with high threshold turn into workers. Consequently, the ratio of workers is maintained at around the equilibrium point determined by the increasing rate $\delta$.

## 3   Simulation-Based Analysis of Response Threshold Model

Through simulation experiments, we evaluate transient behavior during recovery from death of individuals or node failures.

**Table 1.** Parameter setting

| Notation | Description | Default |
|----------|-------------|---------|
| $\theta$ | Hesitation to become a worker | 10 |
| $p$ | Probability of quitting | 0.1 |
| $I_c$ | Interval between two successive stimulus diffusion | 1 |
| $I_d$ | Interval between two successive data reporting | 1 |
| $\delta$ | Increasing rate of stimulus intensity | 0.15 |
| $M$ | Total number of nodes | 1,000 |

## 3.1 Simulation Model

We assume a hypothetical system consisting of a central unit and nodes among which a task is assigned. Nodes are homogeneous and capable of performing the task. The stimulus intensity is diffused to nodes from a central unit. Then as a response to the stimulus each node which received the stimulus decides whether to perform the task or not and reports the decision to the central unit. Those nodes that cannot receive the stimulus do not change their state. The central unit then obtains the number of workers $w(k)$ from received reports and derives the new stimulus intensity by using Eq. (1). The stimulus intensity is again diffused to nodes. While doing a task, a node would inform a central unit of its result. For example, in a case of a monitoring application of a wireless sensor network, each sensor node periodically sends a message containing sensor reading, such as temperature and humidity, to a sink. To take into account a realistic application of the response threshold model, we further assume that exchanges of the stimulus and responses is performed per $I_c$ $(> 0)$ second, corresponding to the control interval of an application system. Furthermore, the data reporting interval is assumed to be $I_d$ $(0 < I_d \leq I_c)$ second.

In the above mentioned system, there are three occasions for disruption of communication between a central unit and a node. A central unit sometimes fails in receiving a response from a node, which we call worker information, for communication errors. As a result, the stimulus intensity will be derived from the wrong number of workers. A central unit would also fail in receiving data from nodes. On the contrary, a node would fail in receiving what we call stimulus information from the central unit and cannot decide its new state. In the analysis, we denote the probability of loss of information as $q_w$ $(0 \leq q_w \leq 1)$, $q_d$ $(0 \leq q_d \leq 1)$ and $q_s$ $(0 \leq q_s \leq 1)$ for worker information, data, and stimulus, respectively, and we call them information loss rate.

## 3.2 Simulation Setting

Based on the hypothetical system we defined in the above, we evaluate the influence of information loss on transient properties during recovery from node failures. There is one task and there are $M$ nodes capable of performing the task. We consider the loss rate is identical among information, that is, $q_w = q_s = q_d = q$. There is no delay in communication. Parameters are summarized in Table 1.

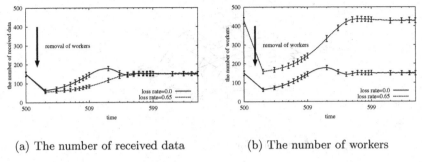

(a) The number of received data      (b) The number of workers

**Fig. 1.** Influence of information loss during recovery from node failure

$\delta = 0.15$ means that a central unit wants to receive as many as 150 data from nodes every $I_d$ seconds. Each simulation run lasts for 1,000 seconds. The fraction $r_f$ $(0 \le r_f \le 1)$ of workers, i.e. nodes with $X_i(k) = 1$, are randomly selected and removed at 500.5 seconds, but $M$ is unchanged reflecting that a central unit is not aware of node removal. In the following, we show averages of 100 simulation runs.

### 3.3 Simulation Evaluation

Figure 1 shows temporal changes of the number of data received by a central unit and the number of workers in the network for the information loss rates of 0.0 and 0.65. We show averages and standard deviation. As shown in Fig. 1(a), independently of the information loss rate, a central unit eventually resumes receiving the desired number of data after node failures. Regarding the transient behavior, the number of received data for higher loss rate slowly increases, while one for no loss increases faster and even overshoots. In this simulation model, the loss of stimulus disturbs appropriate state setting of nodes and as a result higher loss rate leads to slower dynamics. On the contrary, when there is no loss of information, a node can receive a stimulus and change its state appropriately. As a result, the number of workers rapidly increases. However, because of hysteresis in the stimulus calculation, it once overshoots and then reaches the equilibrium.

Figure 1(b) shows the number of workers is larger with the larger information loss rate. When worker information is frequently lost, the stimulus becomes larger than the case without information loss to maintain the underestimated number of workers around 150. Since the loss rate of data is identical to that of worker information, the number of data that a central unit is kept around 150 as show in Fig. 1(a).

Next we evaluate the recovery time $T_r$, which is defined as time required for the number of received data to continuously exceed 96% of the targeted value, which is derived as $\delta \cdot M$, after node failures. In Fig. 2, we depict $T_r$ against the information loss rate $q$ from 0.0 to 0.5 and the node failure rate $r_f$ from 0.0 to

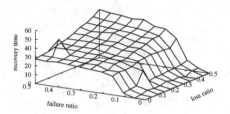

**Fig. 2.** Influence of information loss on the recovery time

0.5. As shown in Fig. 2, the recovery time increases as the rate of information loss increases. For example, the recovery time for $q = 0.0$ and $r_f = 0.5$ is 28 seconds. When the information loss rate is 0.5, the recovery time becomes 54 seconds. Therefore, the increase is as much as double. In the next section, through mathematical analysis, we investigate the influence of control parameters on the recovery time to have faster convergence.

# 4    Mathematical Analysis of Response Threshold Model

From simulation-based analysis, it is shown that the number of workers increases as the loss rate becomes higher. To clarify the range of the loss rate for the response threshold model to be effective, we derive existence conditions of an equilibrium state. We also formulate the recovery time and clarify parameters important to shorten the recovery time.

## 4.1    Analytical Model Considering Information Loss

As explained in section 2, the number of workers in a system returns to the target value after removal of workers by involving non-workers with the increased stimulus. However Fig. 1 shows that there are two types of transient behavior. When a system overreacts to the decrease in the number of workers, it often results in overshooting and further leads to oscillation as illustrated in the left of Fig. 3. Even in that case, it eventually approaches to the target value or the equilibrium state as shown in the left figure. On the contrary, with moderate adaptation, the number of workers steadily increases toward the desired level without oscillation, while it would take longer than the case of aggressive adaptation as shown in the right of Fig. 3.

An information network is a discrete system, where a node intermittently emits and receives a chunk of data called packet and performs predetermined algorithms. However in this paper, for simplicity of analysis, we adopt continuous-time modeling and analyze the effect of information loss on transient behavior, i.e. magnitude of oscillation during recovery from node failure and recovery time

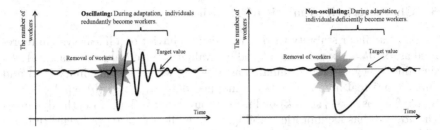

**Fig. 3.** Transient behavior with/without oscillation

of the response threshold model. The validity of analysis will be verified through comparison with simulation results.

Considering loss and reception of information as Bernoulli trials, the expected number of workers and non-workers which receive the stimulus information are formulated as $(1-q_s)n_w$ and $(1-q_s)(M-D-n_w)$, respectively. Here, variable $n_w$ is the expected number of workers in a colony. Constant $D$ $(0 \leq D \leq M)$ is the number of dead workers, assuming that a set of individuals are statically removed from a colony. Then, the expected number $n_s$ of the worker information that a central unit receives from workers is formulated as $(1 - q_w)n_w$. Consequently, the temporal dynamics of the expected value $s$ of the stimulus intensity and the expected number $n_w$ of workers can be formulated as follows.

$$\begin{bmatrix} \frac{ds}{dt} \\ \frac{dn_w}{dt} \end{bmatrix} = \begin{bmatrix} \delta - \frac{(1-q_w)n_s}{M} \\ -p(1-q_s)\,n_w + \frac{s^2}{s^2+\theta^2}(1-q_s)(M-D-n_w) \end{bmatrix} \quad (4)$$

Here we assume that threshold $\theta$ is identical among individuals.

## 4.2  Characteristic Analysis of Equilibrium State

From Eq. (4), we derive the equilibrium state $[\bar{s}\ \bar{n}_w]^T$, where the time variation of the expected values $s$ and $n_w$ are zero.

$$\begin{bmatrix} \bar{s} \\ \bar{n}_w \end{bmatrix} = \begin{bmatrix} \theta\sqrt{\dfrac{p\delta}{(1-\frac{D}{M})(1-q_w)-\delta(1+p)}} \\ \dfrac{\delta M}{1-q_w} \end{bmatrix} \quad (5)$$

Since the stimulus intensity is a positive real number, $(1 - D/M)(1 - q_w) - \delta(1 + p) > 0$ must hold. At the same time, $\delta M/(1 - q_w) < M - D$ must hold so that the number of workers is smaller than the population. Therefore, the condition that a feasible equilibrium state exists is given by the following inequality.

$$(1 - f)(1 - q_w) - \delta(1 + p) > 0 \quad (6)$$

$f$ is $D/M$, that is the ratio of the number of dead individuals to the original colony size.

## 4.3   Characteristic Analysis of Transient State

Based on the linear stability theory, transient behavior of a linear system represented as $d\mathbf{x}/dt = A\mathbf{x}$ can be analyzed by evaluating eigenvalues of matrix $A$. An eigenvalue $\lambda_i$ is generally formulated as $\alpha_i + j\beta_i$, where $j$ is imaginary $\sqrt{-1}$ and $\alpha_i$ and $\beta_i$ are real numbers. $x_i(k)$ is mapped onto $z_i(k) = z_i(0)\exp^{(\alpha_i+j\beta_i)k} = z_i(0)\exp^{\alpha_i k}(\cos\beta_i k + j\sin\beta_i k)$ by linearly converting $\mathbf{x}$. Therefore, the dynamics with $\forall_i \alpha_i < 0$ has asymptotic stability, and smaller $\alpha_i$ leads to longer time for $dz_i/dt$ to converge to 0. In addition, the dynamics with $\forall_i \beta_i = 0$ is stable. $\alpha_i$ is specifically called *damping factor* in this paper.

Linearizing the nonlinear analytical model defined by Eq. (4), we can analyze influence of information loss on transient behavior during recovery from node failures. Firstly, we derive the Jacobian matrix from the nonlinear analytical model. We define $r$ as the fraction of workers in a colony, i.e. $r = n_w/M$. Then, dynamics $ds/dt$ and $dr/dt$ can be formulated as follows.

$$\begin{bmatrix} \frac{ds}{dt} \\ \frac{dr}{dt} \end{bmatrix} = \begin{bmatrix} \delta - (1-q_w)r \\ -p(1-q_s)r + \frac{s^2}{s^2+\theta^2}(1-q_s)(1-f-r) \end{bmatrix} \quad (7)$$

To adopt the linear stability theory, we linearize Eq. (7) at the equilibrium state $[\bar{s}\ \bar{r}]^T = [\theta\sqrt{\frac{p\delta}{(1-f)(1-q_w)-\delta(1+p)}}\ \frac{\delta}{1-q_w}]^T$ by Taylor expansion. Then, the dynamics of error $\mathbf{e} = [e_s\ e_r]^T$ between the equilibrium state $[\bar{s}\ \bar{r}]^T$ and the state $[s\ r]^T$, i.e. $\mathbf{e} = [s\ r]^T - [\bar{s}\ \bar{r}]^T$, can be formulated as the basic linear equation $d\mathbf{e}/dt = A\mathbf{e}$ as follows, where $A$ is the Jacobian matrix.

$$\begin{bmatrix} \frac{de_s}{dt} \\ \frac{de_r}{dt} \end{bmatrix} = \begin{bmatrix} 0 & -(1-q_w) \\ (1-q_s)(1-f-\bar{r})\frac{2\bar{s}\theta^2}{(\bar{s}^2+\theta^2)^2} & -(1-q_s)\left(p+\frac{\bar{s}^2}{\bar{s}^2+\theta^2}\right) \end{bmatrix} \begin{bmatrix} e_s \\ e_r \end{bmatrix} \quad (8)$$

Next, we derive eigenvalues of the matrix $A$ from a characteristic equation $\det|A - \lambda I| = 0$. Since the matrix $A$ in Eq. (8) is in the form of $[0\ a;\ b\ c]$, an eigenvalue is formulated as $0.5(c \pm \sqrt{c^2 + 4ab})$. Specifically, $c$ and $ab$ are formulated as follows, by substituting $\bar{s}$ of Eq. (5).

$$\begin{cases} c = -(1-q_s)p\left\{1 + \frac{\delta}{(1-f)(1-q_w)-\delta}\right\} \\ ab = -\frac{2}{\theta}\sqrt{p\delta}(1-q_s)\frac{\{(1-f)(1-q_w)-\delta(1+p)\}^{1.5}}{(1-f)(1-q_w)-\delta} \end{cases} \quad (9)$$

$c < 0$ and $ab < 0$ must hold for its real part to be always less than 0. When Eq. (6) is satisfied and the equilibrium state exists, $0 < 1-q_s$, $0 < p$, and $\delta < \delta(1+p) < (1-f)(1-q_w)$ hold. Therefore, the first condition $c < 0$ is always met. Similarly, because of $0 < (1-q_w)$, $0 < (1-q_s)$, $0 < \frac{2\bar{s}\theta^2}{(\bar{s}^2+\theta^2)^2}$, and $0 < \bar{r}(1+p) < 1-f-\bar{r}$, the second condition $ab < 0$ is also met. Therefore a real part of the eigenvalue is always negative and as a result the state does not diffuse out of the proximity of the equilibrium state. This means that the

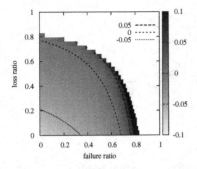 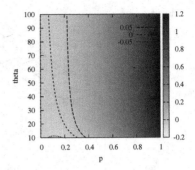

**Fig. 4.** Influence of loss and failure on $c^2 + 4ab$

**Fig. 5.** Influence of parameters $p$ and $\theta$ on $c^2 + 4ab$

response threshold model is robust against information loss once it reaches the equilibrium state.

In the following, we discuss the influence of parameters on transient behavior. For the sake of simplicity, we assume the identical loss rate, i.e. $q_w = q_d = q_s = q$.

**Oscillating or Stable Dynamics.** The dynamics of the state oscillates as shown in the left of Fig. 3 when eigenvalues of the state transition matrix have an imaginary part. In other words, a system reaches the equilibrium state without oscillation when $c^2 + 4ab$ is positive (Fig. 3 right). On the contrary, a system wanders toward the equilibrium state when $c^2 + 4ab$ is negative (Fig. 3 left).

Figure 4 illustrates a value of $c^2 + 4ab$ as a function of the information loss rate and the failure rate. Parameters are set as $p = 0.1$, $\theta = 10$, and $\delta = 0.15$. In the figure, the equilibrium state does not exist in the white area derived from Eq. (6) and we show contour lines of $c^2 + 4ab = 0.05$, 0, and -0.05. A point with a lighter color has a smaller $c^2 + 4ab$. The figure indicates the range that network control based on the response threshold model is feasible. The border can be formulated as $(1-f)(1-q_w) - \delta(1+p) = 0$. Therefore, the range can be extended by choosing appropriate control parameters $\delta$ and $p$ depending on the operational condition expressed by $f$ and $q_w$. In the range, $c^2 + 4ab$ becomes positive and a system steadily moves to the equilibrium when both of the information loss rate and the failure rate are large. However, $c^2 + 4ab$ is negative in the most part. It means that the transient dynamics toward the equilibrium state basically oscillates. The oscillating dynamics causes the overshoot and redundant numbers of nodes temporarily become workers. Whereas a system finally reaches the equilibrium, it becomes a problem in a resource-limited network.

Figure 5 illustrates the influence of tunable parameters, i.e. $p$ and $\theta$ on a value of $c^2 + 4ab$ when information loss rate $q$, failure rate $r_f$, increase rate $\delta$ are set at 0.1, 0.1, and 0.15, respectively. From the figure, we can find that threshold $\theta$ has a small influence on $c^2 + 4ab$. On the contrary, a larger $p$ can make $c^2 + 4ab$

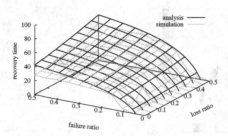

**Fig. 6.** Comparison of analytical simulation results

positive leading to the stable dynamics. However, a larger $p$ makes the second term of left side of Eq. (6) smaller. As a result, loss rate and failure rate need to be larger for the equilibrium state to exist. This implies that the robustness deteriorates as $p$ is large. When we adopt small $p$ to avoid loss of robustness, a system is more likely to oscillate in the transient behavior. However, from a viewpoint of a central unit, receiving a sufficient amount of data is helpful while it is intermittent. Suppressing the degree of oscillation and shortening the duration remain future work.

**Recovery Time from Death.** Recovery time from node failures is defined as necessary time for deviation $\Delta = \bar{n}_w - n_w$ to become as small as 4% of $\bar{n}_w$. Deviation $\Delta(0)$ soon after node failures is $r_f \cdot \bar{n}_w$. The deviation decreases as $\Delta(t) = \Delta(0) \exp^{\alpha_{max} t}$, where $\alpha_{max}$ ($< 0$) is a damping factor derived as $\max Re \frac{c \pm \sqrt{c^2 + 4ab}}{2}$.

Solving the equation $0.04 \cdot \bar{n}_w = \Delta(0) \exp^{\alpha_{max} T_r}$, we can derive recovery time $T_r$ as follows.

$$T_r = \frac{\log_e \frac{0.04}{r_f}}{\alpha_{max}} \tag{10}$$

Figure 6 shows the recovery time derived from Eq. (10) and obtained from simulation results. Although surfaces do not match well between analysis and simulation, the analytical result represents the relationship among failure rate, loss rate, and recovery time. Figure 7 illustrates dependence of the recovery time on these rates for the wider range. As shown in the figure, the maximum recovery time is only double of the minimum. This result supports the robustness of the response threshold model considering that loss often has an exponential influence on conventional control.

Finally, Fig. 8 illustrates the influence of tunable parameters, i.e. quitting probability $p$ and threshold $\theta$, on the recovery time derived from Eqs. (9) and (10), when information loss rate $q$, failure rate $r_f$, and increasing rate $\delta$ are set

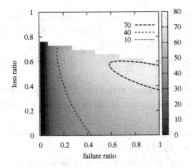

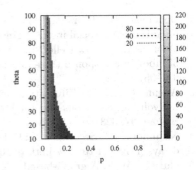

**Fig. 7.** Influence of loss and failure on recovery time

**Fig. 8.** Influence of parameters $p$ and $\theta$ on recovery time

at 0.0, 0.5, and 0.15, respectively. In the white area, the equilibrium state does not exist. The figure shows that threshold $\theta$ does not have much influence on the recovery time for the fixed quitting probability $p$, but a smaller $\theta$ enables wider range of adaptation of quitting probability $p$. To make the recovery time shorter, we should have a small $\theta$ and a large $p$. When $\theta$ is 10, the recovery time decreases from 42.3 to 21.1 by changing quitting probability from 0.1, corresponding to the largest recovery time in Fig. 2, to 0.2.

## 5    Conclusion and Future Work

In this paper, we investigate transient behavior during recovery from failures of individuals in the lossy environment. Results show that the response threshold model is robust against failure and loss. We further analytically clarify the influence of the environmental condition, i.e. loss rate $q$ and failure rate $r_f$, and the control parameters, i.e. threshold $\theta$ and quitting probability $p$, on the oscillation and the recovery time.

As future work, we plan to consider suppression of oscillation in the transient phase to have more stable and faster convergence from failures. Furthermore, we also need to investigate the distribution of the number of workers and the number of received data, because we only consider expected values and equilibrium state in the paper.

**Acknowledgement.** This research was supported in part by Grant-in-Aid for JSPS Fellows of Japan Society for the Promotion of Science (JSPS) and International Collaborative Research Grant of the National Institute of Information and Communications Technology, Japan.

# References

1. Bonabeau, E., Theraulaz, G., Deneubourg, J.-L., Aron, S., Camazine, S.: Self-organization in social insects. Trends in Ecology and Evolution 12, 188–193 (1997)
2. Duarte, A., Pen, I., Keller, L., Weissing, F.J.: Evolution of self-organized division of labor in a response threshold model. Behavioral Ecology and Sociobiology 66, 947–957 (2012)
3. Auchmuty, J.F.G.: Bifurcation analysis of nonlinear reaction-diffusion equations I. Evolution equations and the steady state solutions. Bullebtin of Mathematical Biology 37, 323–365 (1975)
4. Nishimura, J., Friedman, E.J.: Robust convergence in pulse-coupled oscillators with delays. Physical Review Letters 106, 194101 (2011)
5. Gutjahr, W.J.: A graph-based ant system and its convergence. Future Generation Computer Systems 16, 873–888 (2000)
6. Dressler, F., Akan, O.B.: A survey on bio-inspired networking. Computer Networks 54, 881–900 (2010)
7. Meisel, M., Pappasand, V., Zhang, L.: A taxonomy of biologically inspired research in computer networking. Computer Networks 54, 901–916 (2010)
8. Tyrrell, A., Auer, G., Bettstetter, C.: On the accuracy of firefly synchronization with delays. In: Proceedings of the International Symposium on Applied Sciences on Biomedical and Communication Technologies, pp. 1–5 (October 2008)
9. Hyodo, K., Wakamiya, N., Nakaguchi, E., Murata, M., Kubo, Y., Yanagihara, K.: Reaction-diffusion based autonomous control of wireless sensor networks. International Journal of Sensor Networks 7, 189–198 (2010)
10. Bonabeau, E., Sobkowski, A., Theraulaz, G., Deneubourg, J.L.: Adaptive task allocation inspired by a model of division of labor in social insects. In: Proceedings of the International Conference on Biocomputing and Emergent Computation, pp. 36–45 (January 1997)
11. Labella, T., Dressler, F.: A bio-inspired architecture for division of labour in SANETs. Advances in Biologically Inspired Information Systems, 211–230 (December 2007)
12. Janacik, P., Heimfarth, T., Rammig, F.: Emergent topology control based on division of labour in ants. In: Proceedings of the International Conference on Advanced Information Networking and Applications, pp. 733–740 (April 2006)
13. Bonabeau, E., Henaux, F., Guérin, S., Snyers, D., Kuntz, P., Theraulaz, G.: Routing in telecommunications networks with ant-like agents. In: Albayrak, Ş., Garijo, F.J. (eds.) IATA 1998. LNCS (LNAI), vol. 1437, pp. 60–71. Springer, Heidelberg (1998)
14. Sasabe, M., Wakamiya, N., Murata, M., Miyahara, H.: Media streaming on P2P networks with bio-inspired cache replacement algorithm. In: Proceedings of the International Workshop on Biologically Inspired Approaches to Advanced Information Technology, pp. 380–395 (May 2004)

# Emergence of Global Speed Patterns in a Traffic Scenario[*]

Richard Holzer[1], Hermann de Meer[1], and Cristina Beltran Ruiz[2]

[1] Faculty of Informatics and Mathematics, University of Passau,
Innstraße 43, 94032 Passau, Germany
{holzer,demeer}@fim.uni-passau.de
[2] SICE, Sepulveda 6, E-28108 Alcobendas, Madrid, Spain
cbeltran@sice.com

**Abstract.** We investigate different analysis methods for traffic data.
The measure for emergence can be used to identify global dependencies
in data sets. The measure for target orientation can be used to identify
dangerous situations in traffic. We apply these measures in a use case
on a data set of the M30 highway in Madrid. The evaluation shows
that the measures can be used to predict or to identify abnormal events
like accidents in traffic by an evaluation of velocity data or flow data
measured by detectors at the road. Such events leads to a decrease of
the measures of emergence and target orientation.

**Keywords:** Quantitative Measures, Emergence, Target Orientation,
Traffic Safety.

## 1 Introduction

In traffic, abnormal events like accidents on highways may cause a sudden drop
in traffic flow, which might lead to dangerous situations, that may result in
follow-on accidents [1]. To increase safety in traffic, Vehicle-to-vehicle (V2V)
and vehicle-to-infrastructure (V2I) communication technologies could be used to
provide other drivers with the necessary information or to give a direct advice
(e.g. "slow down") to avoid further accidents. In [2] some V2V systems have been
proposed to achieve this goal. To be able to advice drivers to a specified behavior,
an evaluation of the current situation is necessary: How can a dangerous situation
be predicted or at least identified by using measurements of data? Measurements
of the parameters of the vehicles (e.g. position or speed) can either be done
by using a system inside of the vehicles or by a system in the infrastructure.
Although a system in the infrastructure has no access to the whole trajectory of
a vehicle, a set of sensors at the road can measure the parameters of the vehicles
passing the sensors. If we consider a set of such detectors at different positions,

---

[*] This research is partially supported by the SOCIONICAL project (FP7, ICT-2007-
3-231288), by the Network of Excellence EINS (FP7, ICT-2011-7-288021) and by
the Network of Excellence EuroNF (FP7, ICT-2007-1-216366).

W. Elmenreich, F. Dressler, and V. Loreto (Eds.): IWSOS 2013, LNCS 8221, pp. 39–53, 2014.
© IFIP International Federation for Information Processing 2014

the data of all detectors might be sufficient to identify abnormal events (e.g. an accident) in the traffic flow. For this purpose, measures are needed to quantify the safety relevant properties of the current state.

In the recent years, some quantitative measures for properties in self-organizing systems have been developed [3], [4], [5]:

- Autonomy
- Emergence
- Global state awareness
- Target orientation
- Resilience
- Adaptivity

In a traffic scenario, some of these measures can be used for the analysis of safety relevant properties in the system: By specifying the overall goal of the system in form of a fitness function, the measures for target orientation, resilience and adaptivity can be used in different contexts to evaluate the average fitness of the system and to identify abnormal events like accidents. Also the measure of emergence might be useful for the interpretation of the measured data: A high emergence of velocities indicates high dependencies between the velocities of different vehicles, while a low emergence of velocities indicates few dependencies, which can be interpreted as a disharmony in the traffic flow.

After the identification of the presence of an abnormal event, the harmonization of the traffic flow (see [6] and [2]) can be increased either by overhead speed signs or direct signals between the vehicles (in case of V2V communication).

In this paper we analyze a set of 349920 data collected and received from detectors at the highway M30 in Madrid. M30 highway is controlled by an Intelligent Transport System (ITS) implemented as a four-level control architecture that comprises two control centers (main and backup) at the top level, the intermediate level of communications, the distributed intelligence level and field equipment at the bottom level; the latter is responsible of continuously measuring different traffic values and of producing output values at different time intervals, normally minutes.

For this study, access to a specific set of registered data measured by the ITS system implemented in M30 has been granted between SICE as co-author of the paper and Madrid City Council as owner of the data. The collected dataset used in this study include output values of velocity, intensity and occupancy measured by minute at each lane of a road segment of around 12 km of the whole highway. Those data are subject to IPR protection and will not be reproduced in this paper.

The measure of emergence [3] applied to the velocity data can be used to measure dependencies between the velocities of different cars. Analogously the emergence measure can also be applied to flow data (temporal density) to detect dependencies in flow data. The measure for target orientation [4] shows, whether the overall goal (traffic safety) is satisfied. The goal is formally specified by using the variance coefficient of velocities (see [7] and [2]). Section 2 specifies the micro-level model of systems based on [3]. Sections 3 and 4 specify the methodology for the evaluation of emergence and target orientation in systems. In Section 5 these measures are applied to a data set of the M30 highway in Madrid. In Section 6 the results of the evaluation are discussed and Section 7 concludes the paper.

## 2   Discrete Micro-level Model of Systems

We consider a discrete system $S$ with many entities and interactions between the entities. If the topology of such a system is static then it can be represented [3] by a directed graph $G = (V, E)$, where the nodes of the graph represent the entities and the edges represent the communication channels. For dynamic topologies a time dependent directed graph can be used. Each entity has an internal state, which might change during the time. The communication channels between the entities can be used for interaction between the entities, i.e. an entity can send some data to some other entities, which may have some influence on their internal states. For the set of all points in time, when a change of the system occurs (e.g. a change in the state of an entity or an interaction between entities), we use $T = \mathbb{N}_0 = \{0, 1, 2, \ldots\}$.

Let $Conf$ be the set of all global configurations of $S$, i.e. each $c = (c_V, c_E) \in Conf$ is a pair, where $c_V = (c_v)_{v \in V}$ contains the current internal state $c_v$ for each entity $v \in V$ and $c_E = (c_e)_{e \in E}$ contains the current value $c_e$ on the communication channel for each $e \in E$. The behavior of the entities (change of the state and interaction) might be deterministic or stochastic. Also the initialization of the system at time $t = 0$ might be deterministic or stochastic. For $t \in T$ let $Conf_t$ be the random variable for the configuration at time $t$ taking values in $Conf$. A trajectory of the system is a realization $(c_0, c_1, c_2, \ldots) \in Conf^T$ of the stochastic process $(Conf_t)_{t \in T}$. The trajectory describes the sequence of configurations during a run of the system: At each time $t \in T$ the system is in a configuration $c_t \in Conf$, and in the next step the changes in the entities and in the communication channels lead to the next configuration $c_{t+1} \in Conf$ at time $t + 1$.

## 3   Emergence

In [3] the measure of emergence at each point in time is defined by using the entropies of the communication channels between the entities. Here we generalize this concept to an arbitrary family of measures defined on the set of global states.

For a discrete random variable $X : \Omega \to W$ on a probability space $\Omega$ into a set $W$ of values the entropy [8] is defined by

$$H(X) = - \sum_{w \in W} P(X = w) \log_2 P(X = w)$$

with $0 \cdot \log_2 0 := 0$. This entropy measures the average number of bits needed to encode the outcome of the random variable $X$ in an optimal way. If $X$ has a uniform distribution on a finite set $W$, then we get $H(X) = \log_2(|W|)$. For other distributions of $X$ this value is an upper bound, i.e. we always have $H(X) \leq \log_2(|W|)$. The entropy measure can also be applied on tuples of random variables:

$$H(X,Y) = - \sum_{w \in W, w' \in W'} P(X = w, Y = w') \log_2 P(X = w, Y = w')$$

The entropy of a pair $(X, Y)$ of random variables is not greater than the sum of the entropies of the random variables:

$$H(X,Y) \leq H(X) + H(Y)$$

To formally define the level of emergence of the system, we have to specify the properties of interest. These properties of interest can be specified by a family of maps. Let $J$ be an at most countable set and $m = (m_j)_{j \in J}$ be a family of maps

$$m_j : Conf \to W_j$$

for a value set $W_j$ for $j \in J$. For $c \in Conf$ let

$$m(c) = (m_j(c))_{j \in J} \in \prod_{j \in J} W_j$$

The level of emergence at time $t \in T$ is defined by

$$\varepsilon_t = 1 - \frac{H(m(Conf_t))}{\sum\limits_{j \in J} H(m_j(Conf_t))}$$

Note that we have

$$\varepsilon_t \in [0, 1]$$

because of

$$H(m(Conf_t)) \leq \sum_{j \in J} H(m_j(Conf_t))$$

Therefore we get a map

$$\varepsilon : T \to [0,1] \quad \text{with} \quad t \mapsto \varepsilon_t$$

Note that the definition of level of emergence in [3] is a special case of this definition: We can choose $J = E$ as the set of all communication channels and $m_e(c_V, c_E) = c_e$ as the value on the channel $e \in E$. While [3] only considers emergence in communication, the definition above can be used in a much broader sense. First, the measures $m = (m_j)_{j \in J}$ have to be specified to describe, in which type of data we would like to identify emergent patterns. Then the global emerging patterns are measured by comparing the entropy of the specified measures $m$ in the current configuration with the sum of the entropies

$H(m_j(Conf_t))$ for $j \in J$. If the values of the measures $m_j$ are independent, then the whole entropy is identical to the sum of the entropies:

$$H(m(Conf_t)) = \sum_{j \in J} H(m_j(Conf_t))$$

In this case we get $\varepsilon_t = 0$. The more dependencies are between the measures $m_j$, the smaller is the global entropy $H(m(Conf_t))$, i.e. the quotient

$$\frac{H(m(Conf_t))}{\sum\limits_{j \in J} H(m_j(Conf_t))}$$

decreases with higher dependencies. Therefore the value $\varepsilon_t$ indicates the existence of dependencies (which can be seen as global patterns) in the values of the measures $m_j$: If the emergence $\varepsilon_t$ is near 0, then all measures $m_j$ are nearly independent at the current point in time $t \in T$. If the emergence $\varepsilon_t$ is near 1, then there are many dependencies between the measures at the current point in time.

An analytical evaluation of the level of emergence in a complex system is usually impossible because of the huge state space $Conf$. Therefore we need approximation methods [9] to be able to evaluate at least an approximation of the emergence of a system.

To transfer the definition of the entropy $H(X)$ of a random variable $X$ to a sequence $(X(\omega_1), \ldots, X(\omega_n))$ of realizations of the random variable $X$, we use relative frequencies of the realizations for the probabilities. This leads to the following definition: For a tuple $x = (x_1, \ldots, x_n) \in W^n$ the relative frequency of a value $w \in W$ in $x$ is defined by

$$rel_w(x) := \frac{|\{i \leq n \ : \ x_i = w\}|}{n}$$

The entropy $H(x)$ of $x$ is defined by

$$H(x) = -\sum_{w \in W} rel_w(x) \log_2 rel_w(x)$$

This definition can be extended to matrices[1] $A = (A_{ij})_{i \leq n, j \leq r} \in W^{n \times r}$ by identifying $A$ with the sequence of rows of $A$. For $i \leq n$ and $j \leq r$ let $A_{i\cdot} \in W^r$ be the $i$-th row of $A$ and let $A_{\cdot j} \in W^n$ be the $j$-th column of $A$. The relative frequency of a tuple $w = (w_1, \ldots, w_r) \in W^r$ in $A$ is

$$rel_w(A) := \frac{|\{i \leq n \ : \ A_{i\cdot} = w\}|}{n}$$

---

[1] It is also possible to use "inhomogeneous" matrices, where the entries of different columns come from different value sets $W_j$. In this case, the matrix $A$ can be seen either as an element of $\prod\limits_{j \leq r} W_j^n$ (sequence of columns) or as an element of $\left( \prod\limits_{j \leq r} W_j \right)^n$ (sequence of rows).

The entropy of $A$ is defined by

$$H(A) = - \sum_{w \in W^r} rel_w(A) \log_2 rel_w(A)$$

The emergence of $A$ is defined by

$$\varepsilon(A) = 1 - \frac{H(A)}{\sum\limits_{j \leq r} H(A._j)}$$

Now we can use these concepts to approximate the level of emergence $\varepsilon_t$ of a large system at time $t \in T$ with respect to a given family $m = (m_j)_{j \in J}$ of measures with a finite set $J$. For this purpose we need some realizations of the stochastic process $(Conf_t)_{t \in T}$. Such realizations can either be received from simulations or from measured real data. Let $n$ be the number of realizations and for $1 \leq i \leq n$ let $Conf_{t,i} \in Conf$ be the global state at time $t \in T$ for the $i$-th realization. For each $t \in T$ we can define the $n \times |J|$ matrix $A = (A_{ij})_{i \leq n, j \in J}$ by $A_{ij} = m_j(Conf_{t,i})$. Then we can observe:

– For $i \leq n$ the $i$-th row of this matrix is the $i$-th realization of $m(Conf_t)$:

$$A_{i.} = m(Conf_{t,i})$$

– For $j \in J$ the $j$-th column of $A$ consists of all realizations of $m_j(Conf_t)$:

$$A._j = (m_j(Conf_{t,i}))_{i \leq n}$$

Then we can use $\varepsilon(A)$ as an approximation of $\varepsilon_t$: The relative frequencies of the values $w = m_j(Conf_{t,i})$ correspond to the probabilities $P(m_j(Conf_t) = w)$ and the relative frequencies of the tuples

$$w = (w_1, \ldots, w_{|J|}) = m(Conf_{t,i}) = A_{i.}$$

correspond to the probabilities $P(m(Conf_t) = w)$. In Section 5 this approximation $\varepsilon_t \approx \varepsilon(A)$ will be used for the evaluation of an approximation of the emergence in a traffic scenario by using real data.

## 4    Target Orientation

The goal of a system [4] can be specified by a fitness function $f : Conf \to [0, 1]$ on the global state space. The fitness function describes, which configurations are good ($f(c) \approx 1$) and which configurations are bad ($f(c) \approx 0$). At each point in time $t \in T$ this fitness function can be applied to $Conf_t$ to measure the fitness of the system at time $t$. Since $Conf_t$ is a random variable, $f(Conf_t)$ is also a random variable. The level of target orientation of the system at time $t \in T$ is the mean value of this random variable:

$$TO_t = E(f(Conf_t))$$

This induces a map

$$TO : T \rightarrow [0,1] \quad \text{with} \quad t \mapsto TO_t$$

As for the measure for the emergence, the measure of target orientation usually can not be calculated analytically in large systems, so we need again realizations (either by simulations or real data) for the stochastic process $(Conf_t)_{t \in T}$. Let $n$ be the number of realizations, $t \in T$ and for $1 \leq i \leq n$ let $Conf_{t,i}$ be the $i$-th realization of the random variable $Conf_t$. Then the mean value $E(f(Conf_t))$ of the fitness function can be approximated by the average fitness of the realizations:

$$TO_t \approx \frac{1}{n} \sum_{i=1}^{n} f(Conf_{t,i})$$

The accuracy of this approximation increases with the number $n$ of realizations, but in some cases it is not possible to produce many realizations of the random variable $Conf_t$. For example, if the measure of target orientation is applied to real data measured in a given system, then we have only one realization of $Conf_t$ for each point in time $t$. While for the measure of emergence a single realization is not sufficient to calculate an approximation of $\varepsilon_t$, this is not so critical for the level of target orientation: The target orientation $TO : T \rightarrow [0,1]$ can be approximated by applying the fitness function to the realization:

$$TO_t \approx f(Conf_{t,1})$$

For the emergence we will see in the next section, that aggregation techniques can help to overcome this problem.

## 5   Evaluation of Highway Data

In this section we evaluate data measured at the highway M30 in Madrid. We use 81 detectors of the whole M30 road network, each of them located in one road lane, to collect minute by minute measures of the velocity of cars passing the detectors during one working day. Table 1 shows the positions in $km$ (relative to a reference point) of detectors at 24 different positions at the M30 highway. Let $P$ be the set of these 24 positions. For each position $p \in P$ the second row of Table 1 shows the number of lanes $L_p$. For each lane $l \in \{1, 2, \ldots, L_p\}$ a detector at position $p$ measures some data for lane $l$. Once per minute, each detector produces some output values. The output of the detector is (among others)

- the number of vehicles, which have passed the detector at position $p$ in this lane $l$ in a time interval of length $1min$, and
- the average value of the measured velocities of these vehicles, when they passed the detector in the time interval.

**Table 1.** Positions of detectors in $km$ (relative to a reference point) and number of lanes

| Position (km) | 0.0 | 0.2 | 0.9 | 1.4 | 2.1 | 2.4 | 3.4 | 4.0 | 4.4 | 4.7 | 5.0 | 5.6 | 6.1 | 6.6 | 7.1 | 8.3 | 8.6 | 9.0 | 9.4 | 9.8 | 10.7 | 11.0 | 11.6 | 12.0 |
|---|---|---|---|---|---|---|---|---|---|---|---|---|---|---|---|---|---|---|---|---|---|---|---|---|
| Lanes | 3 | 3 | 3 | 3 | 3 | 3 | 3 | 4 | 3 | 3 | 3 | 3 | 4 | 3 | 4 | 5 | 5 | 3 | 3 | 3 | 5 | 2 | 3 | 4 |

We analyze the emergence and the target orientation of the measured data of one day (=1440 minutes), where an accident has occurred on the highway at time $t = 1170min$ (which is 7:30pm on that day) near position $p = 6.1$. For each detector position $p \in P$ and each lane $l \le L_p$ we have some output values of the detectors for the time intervals $I_t := [t, t+1[$ of length $1min$ for $t \in T' := \{0, 1, \ldots, 1439\}$. Let $N_{p,l,t}$ the number of vehicles passing the detector in the time interval $I_t$ and let $v_{p,l,t}$ be the averaged velocity values of these vehicles measured at the detector.

For our micro-level model we can use the set of all detectors

$$D = \{d_{p,l} \mid p \in P, l \le L_p\}$$

as a subset of all entities: $D \subseteq V$. Other entities are the vehicles on the road. The interactions between the vehicles and the detectors cause a change of the internal states of the detectors $d \in D$. In a configuration $c \in Conf$ each detector $d_{p,l} \in D$ contains two values in its internal state: $N_{p,l,c}$ is the number of vehicles detected and $v_{p,l,c}$ is the average measured velocity of these vehicles. We assume a finite accuracy for the detectors, so the set of all possible outputs of a detector $d_{p,l} \in D$ is finite:

$$|\{(N_{p,l,c}, v_{p,l,c}) : c \in Conf\}| < \infty \quad \text{for} \quad p \in P, \quad l \le L_p$$

### 5.1    Emergence of the Data Set

To be able to evaluate the emergence of the data in the traffic scenario, we now have the problem, that we do not have many realizations of the random variable $Conf_t$, so the methodology of section 3 can not be applied directly. But an aggregation of data can help to solve this problem: When we consider the average velocity $v_{p,l,t}$ measured in a time interval $I_t$ at one detector at position $p \in P$ on lane $l \le L_p$, then there will be usually only a small change to the value in the next time interval $I_{t+1}$, i.e. the difference $|v_{p,l,t} - v_{p,l,t+1}|$ is usually not very large. Therefore we can aggregate a number $q \ge 1$ of time intervals $I_t, I_{t+1}, \ldots, I_{t+q-1}$ into one large time interval

$$I_{t,q} := I_t \cup I_{t+1} \cup \ldots \cup I_{t+q-1}$$

and consider the corresponding velocity values

$$v_{p,l,t}, v_{p,l,t+1}, \ldots, v_{p,l,t+q-1}$$

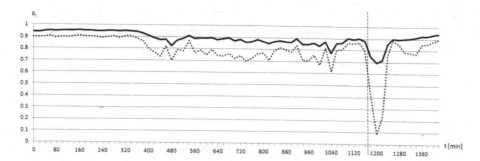

**Fig. 1.** Emergence $\varepsilon$ of the velocity in dependency of the time $[min]$. For the solid line the data of all detectors were used, while for the dotted line only the data of detectors with position $p \in [0km, 6.1km]$ were used. An accident happened near position $6.1km$ at time $t = 1170$.

as different realizations of the random variable $V_{p,l,t,q}$ for the average velocity in $I_{t,q}$ at position $p$ on lane $l$. In the following we use $q = 20$. For each $t \in \{0, 20, 40, \ldots 1420\}$ we have now $q = 20$ realizations of the random variable $V_{p,l,t,q}$. Now we have to fix the measures $m = (m_j)_{j \in J}$ for the emergence of the system. We could use the the measure

$$m_{p,l}(c) = v_{p,l,c} \quad \text{for} \quad p \in P, \quad l \le L_p$$

but this would cause very low values for the relative frequencies, because two realizations of the velocity usually have a different value. We use again the concept of aggregation, but this time on the values for the velocity: We do not consider single velocity values, but intervals $K_0, K_1, \ldots$ of velocity values. For this purpose we fix the interval length $k = 30km/h$ and get the intervals

$$K_s = [s \cdot k, (s+1) \cdot k[ \quad \text{for} \quad s = 0, 1, 2, \ldots$$

The corresponding measure is

$$m_{p,l}(c) = \lfloor v_{p,l,c}/k \rfloor$$

This measure yields the value $s$ such that the realization of the velocity is in the interval $K_s$. As mentioned in Section 3 the values of the measures of the realizations form a matrix $A$, where the $i$-th row of the matrix is $m(Conf_{t,i})$, i.e. the $i$-th row consists of the values $m_{p,l}(Conf_{t,i})$ for all positions $p \in P$ and all lanes $l \le L_p$. The approximated emergence $\varepsilon_t \approx \varepsilon(A)$ in dependency of the time $t \in \{0, 20, 40, \ldots 1420\}$ is shown in Figure 1 (solid line).

At $t = 1170$ there is an accident on the highway near position $p = 6.1$, which causes a blocked lane. After this point in time we see a decrease of the emergence for a duration of about 80 minutes. Then the emergence gets back to a normal value. In a normal traffic flow, the speed of vehicles is much more harmonized than for situations with a blocked lane (caused by an accident or some other abnormal event), so a normal traffic flow shows more dependencies

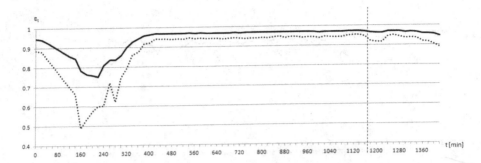

**Fig. 2.** Emergence $\varepsilon$ of the vehicle flow in dependency of the time $[min]$. For the solid line the data of all detectors were used, while for the dotted line only the data of detectors with position $p \in [0km, 6.1km]$ were used. An accident happened near position $6.1km$ at time $t = 1170$.

between the velocities of the vehicles, which is indicated by a higher value for the emergence $\varepsilon_t$. When we restrict our evaluation on the area $[0km, 6.1km]$ (before the accident), then the decrease of the emergence at time $t = 1170$ is much stronger (dotted line in Figure 1). This results from the fact that in the area behind the accident $[6.1km, 12km]$ the traffic flow is much more harmonized than before the accident, so the restriction of the area of interest to the problem zone leads to a stronger irregularity in the graph of $\varepsilon$ compared to the normal traffic flow.

Next to the emergence of the velocity data, we can also investigate the emergence of the flow of vehicles (temporal density). The value $N_{p,l,t}$ measures the flow of cars at position $p \in P$ for lane $l \leq L_p$ with the unit $min^{-1}$. For the evaluation of the emergence of the flow data, we use the same methodology as for the velocity data. The only difference is the aggregation size on the value set: For the velocity data we aggregated the data into intervals of length $k = 30km/h$, while for the flow data we use the interval length $k = 4min^{-1}$. The aggregation on the time axis is for the flow data the same as for the velocity data: $q = 20$. The results are shown in Figure 2. At the time of accident, the evaluation of the whole data set (solid line in Figure 2) shows only a very small decrease. When we only evaluate the data of the detectors before the accident (i.e. with position $p \leq 6.1km$) there is a larger decrease at the time of accident (dotted line in Figure 2). The graph shows that for $t \in [80min, 320min]$ the emergence of the flow is below the normal value. This might result from the low traffic during the night: From 1:20am (which is $t = 80min$) to 5:20am (which is $t = 320min$) there is much less traffic on the road, so a disharmony in the traffic flow is more likely than on a full road. In the time interval $[80, 320[$ the average number of detected cars per minute is

$$N_{avg} = \frac{1}{240} \sum_{t=80}^{319} N_t \approx 102$$

where

$$N_t = \sum_{p \in P, l \leq L_p} N_{p,l,t}$$

is the total number of cars detected in the interval $I_t$ (which is usually much higher than the number of cars on the road segment, because each car is counted by more than one detector in $1min$). In the time interval $[320, 1440[$ the average number of detected cars is $N_{avg} \approx 1357$. While for the emergence of velocities the difference of the amount of traffic had only a small effect on the results (see Fig. 1), the emergence measure for the flow should only be interpreted in connection with the data for the amount of traffic.

## 5.2 Target Orientation of the Data Set

To be able to evaluate the highway system with the measure for target orientation, we have to specify the goal of the system. The main goal is the safety, so we have to define, which configurations are safe and which configurations are dangerous. In [2] three different measures for safety in traffic have been proposed:

- based on the variance of the velocities,
- based on the variance of accelerations,
- based on the mean time to collision (TTC)

From the data of the detectors at the M30 highway we are not able to calculate the accelerations or the TTC values for the vehicles, but we have some aggregated velocity values. Each detector yields the number of vehicles $N_{p,l,t}$ which have passed the detector at position $p \in P$ in lane $l \leq L_p$ in the interval $I_t$. The average measured velocity of these vehicles can be used to specify the fitness function $f : Conf \rightarrow [0, 1]$ on the set of all global configurations. We consider a configuration as dangerous, if the detectors yield velocity data with a high variance coefficient, because in this case an accident is more likely than for a homogeneous traffic flow, where the velocities of the vehicles are very similar. We only consider configurations, where at least one vehicle with positive velocity was detected (otherwise the variance coefficient of velocities does not exist). Let $c \in Conf$. Let

$$N_c = \sum_{p \in P, l \leq L_p} N_{p,l,c}$$

be the total number of detected vehicles for the configuration $c$. The average measured velocity of the detected vehicles in the configuration $c$ is

$$\mu_c = \frac{1}{N_c} \sum_{p \in P, l \leq L_p} N_{p,l,c} \cdot v_{p,l,c}$$

For the empirical variance of the velocities, the detectors do not provide enough information, because some velocities are already aggregated into a single value $v_{p,l,c}$. But we can use $N_{p,l,c}$ as a weight in the formula of the empirical variance:

$$\sigma_c^2 = \frac{1}{N_c - 1} \sum_{p \in P, l \leq L_p} N_{p,l,c} \cdot (v_{p,l,c} - \mu_c)^2$$

Then
$$\frac{\sigma_c}{\mu_c}$$
is the empirical variance coefficient of the velocity for the configuration $c$. Let
$$K \geq \max_{c \in Conf} \frac{\sigma_c}{\mu_c}$$
be a normalization constant.[2] The fitness $f(c)$ of the configuration $c$ is defined by
$$f(c) := 1 - \frac{1}{K} \cdot \frac{\sigma_c}{\mu_c}$$
A low value for the fitness implies a high variance coefficient for the velocities, i.e. the "bad states" are those configurations, where the velocities of different vehicles have a relative high variance.

For $t \in T' := \{0, 1, \ldots, 1439\}$ we have $n = 1$ realization $Conf_{t,1}$ of the random variable $Conf_t$ for the configuration at time $t$ containing all necessary information ($v_{p,l,t}$ and $N_{p,l,t}$) for the calculation of the fitness of the configuration $Conf_{t,1}$. The level of target orientation at time $t \in T'$ is approximated by the fitness of the realization:
$$TO_t \approx f(Conf_{t,1}) = 1 - \frac{1}{K} \cdot \frac{\sigma_t}{\mu_t}$$
where $\frac{\sigma_t}{\mu_t}$ is the empirical variance coefficient of measured velocities in the interval $I_t$ for the realization $Conf_{t,1}$. For our data set we can set $K := 1$ because all variance coefficients are below 1. The approximation of the target orientation $TO : T' \to [0, 1]$ is shown in Figure 3.

At time $t = 1170$ an accident has occurred. In Figure 3 we see that the level of target orientation is low at $t = 1170$, we have $TO_t = 0.68$. This means that we have a high variance coefficient for the velocities. After the accident, the level of target orientation still decreases to its minimum $TO_t = 0.56$ at time $t = 1247$ and then it goes back to a normal level $\geq 0.9$ at time $t = 1291$.

## 6   Discussion of the Results

In the previous section we applied measures on the M30 highway data set:

- emergence of velocity,
- emergence of flow,
- target orientation with respect to the variance coefficient of velocities.

For these measures, an accident on the highway causes a significant decrease of the values of the measures. For the level of emergence in velocities this can be explained by the fact that a homogeneous traffic flow shows more dependencies in the velocity data than a traffic flow that is disturbed by an event like an

---

[2] Such $K$ exists because of the finite accuracy of the detectors.

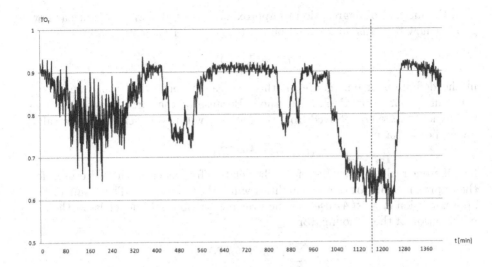

**Fig. 3.** Target orientation $TO$ of the system in dependency of the time $[min]$

accident. Analogously this argument holds for the level of emergence in flow, but the decrease is not so strong like for the emergence of velocities. For the level of target orientation we already have a low value at the point in time, when the accident occurred, so the low fitness value of the global configuration indicates a dangerous situation, where accidents are more likely than for harmonic traffic flows with a low variance coefficient of velocities. Therefore an evaluation of the measure for target orientation can be used as an indicator for the prediction of accidents: A high value for $TO_t$ indicates a low probability for accidents, while a low value for $TO_t$ indicates a higher probability. For the level of emergence of velocities and densities, the values of the measures were high at the point in time, when the accident occurred, so these measure do not predict the accident, but since they decreases after the accident, all three measures can be used as indicators for the identification of accidents: If they have low values, then this might be caused by an abnormal event like an accident.

Also in the situation after the accident the level of target orientation and the levels of emergence stay at low values for more than one hour. Due to the accident, one lane of the highway was blocked, so a traffic jam occurred near the place of accident. The vehicles near the accident have a low velocity, while the vehicles far away from the accident have a high velocity, so the variance of the velocities is high, which is indicated by a low value of target orientation. Concerning the emergence of velocities, this situation shows less dependencies between the velocities of different vehicles than a harmonic traffic flow without a blocked lane, so we get a low value for the emergence for the velocity data.

For the level of emergence we have the problem that we are not able to calculate the exact value $\varepsilon_t$ because of the huge state space of the system, so we used the method of aggregation to approximate the value $\varepsilon_t$. It is unclear how good or how bad this approximation is. If we consider the enumerator

$$H(m(Conf_t))$$

in the definition of emergence $\varepsilon_t$, then we can observe that this value stays constant in our approximation method: Because of the large state space two realizations of $Conf_t$ received from the data set will never yield the same value for $m$, i.e. we have

$$m(Conf_{t,i}) \neq m(Conf_{t,j})$$

for different realizations. Therefore, the relative frequencies, which are used for the approximations of the probabilities will either be 0 or 1. The result of the approximation $\varepsilon_t \approx \varepsilon(A)$ used in the previous section only depends on the approximation of the denominator

$$\sum_{j \in J} H(m_j(Conf_t))$$

in the definition of the emergence. The evaluation shows that the changes in the denominator are sufficient to identify abnormal events like an accident. Therefore we can conclude that the value of the approximation $\varepsilon(A)$ might differ very much from the real value $\varepsilon_t$, but changes in the function $\varepsilon : T \rightarrow [0, 1]$ can also be found in the approximation, so the approximation contains enough information to be able to identify some properties of interest.

## 7   Conclusion and Future Work

In this paper we have investigated different analysis methods for traffic data: Emergence and target orientation. The measure for emergence applied to a data set of velocities or flow data can be used to identify global dependencies in the data set. The measure for target orientation can be used to identify dangerous situations in traffic. We applied both measures in a use case on a data set of the M30 highway in Madrid. The evaluation shows that the measures can be used to identify and predict abnormal events like accidents in traffic by an evaluation of velocity data: A decreasing level emergence of velocities indicates a reduction of dependencies between the velocities of different vehicles, which implies a disharmony in the traffic flow. A decreasing level of target orientation with respect to the specified fitness function indicates a higher variance of velocities, so we again have an indication for the disharmony in the traffic flow, which implies a more dangerous situation.

Note that these measures can not only be applied for the identification of accident but for arbitrary events disturbing the harmony of the traffic flow: Whenever the measures indicate a disharmonious traffic flow, a change of the behavior of the drivers might be desirable to harmonize the flow again and to increase safety in traffic. By using detectors at roads, which measures the velocities of passing vehicles, the evaluations of the measures can be used to control overhead signs for speed limits to increase the safety of the traffic scenario, either by increasing the harmonization of the flow to prevent accidents or — in the case, that already an accident (or any other abnormal event disturbing the traffic flow) has already happened — to prevent follow-on accidents.

For future work it might be interesting to evaluate other (non-highway) data with some other abnormal events (e.g. traffic jams without accidents) disturbing the harmonization of the traffic flow to see, whether the proposed measures can also be used in other scenarios.

# References

1. Masinick, J., Teng, H.: An analysis on the impact of rubbernecking on urban freeway traffic. Tech. Rep., Centre for Transportation Studies of the University of Virginia (2004)
2. Fullerton, M., Holzer, R., De Meer, H., Beltran, C.: Novel assessment of a peer-peer road accident survival system. In: IEEE Self-adaptive and Self-organizing Systems Workshop Eval4SASO 2012 (2012)
3. Holzer, R., de Meer, H., Bettstetter, C.: On autonomy and emergence in self-organizing systems. In: Hummel, K.A., Sterbenz, J.P.G. (eds.) IWSOS 2008. LNCS, vol. 5343, pp. 157–169. Springer, Heidelberg (2008)
4. Holzer, R., de Meer, H.: Quantitative Modeling of Self-organizing Properties. In: Spyropoulos, T., Hummel, K.A. (eds.) IWSOS 2009. LNCS, vol. 5918, pp. 149–161. Springer, Heidelberg (2009)
5. Auer, C., Wüchner, P., de Meer, H.: The degree of global-state awareness in self-organizing systems. In: Spyropoulos, T., Hummel, K.A. (eds.) IWSOS 2009. LNCS, vol. 5918, pp. 125–136. Springer, Heidelberg (2009)
6. Assenmacher, S., Killat, M., Schmidt-Eisenlohr, F., Vortisch, P.: A simulative approach for the identification of possibilities and impacts of v2xcommunication. In: 15th World Congress on ITS, pp. 29–38 (2008)
7. Klein, A.: Untersuchung der Harmonisierungswirkung von Streckenbeeinflussungsanlagen. In: 4. Aachener Simulations Symposium (2011)
8. Cover, T.M., Thomas, J.A.: Elements of Information Theory. 2nd edn. Wiley (2006)
9. Holzer, R., De Meer, H.: Methods for approximations of quantitative measures in self-organizing systems. In: Bettstetter, C., Gershenson, C. (eds.) IWSOS 2011. LNCS, vol. 6557, pp. 1–15. Springer, Heidelberg (2011), The original publication is available at www.springerlink.com

# On the Efficiency of Information-Assisted Search for Parking Space: A Game-Theoretic Approach*

Evangelia Kokolaki, Merkourios Karaliopoulos, and Ioannis Stavrakakis

Department of Informatics and Telecommunications,
National & Kapodistrian University of Athens
Ilissia, 157 84 Athens, Greece
{evako,mkaralio,ioannis}@di.uoa.gr

**Abstract.** This paper seeks to systematically explore the efficiency of the uncoordinated information-assisted parking search in urban environments with two
types of parking resource facilities: inexpensive but limited facilities (public) and
expensive yet unlimited ones (private); an additional cruising cost is incurred
when deciding for a public facility but failing to actually utilize one. Drivers decide whether to go for the public or directly for the private facilities, assuming
perfect knowledge of prices and costs, total parking capacities and demand; the
latter information can be broadcast by an ideal centralized information dissemination mechanism, assisting the otherwise uncoordinated parking search process.
Drivers are viewed as strategic decision-makers that aim at minimizing the cost
of the acquired parking spot. We formulate the resulting game as an instance of
resource selection games and derive its Nash equilibria and their dependence on
the environmental parameters such as the parking demand and supply as well as
the pricing policy. The cost at the equilibrium states is compared to that under
the optimal resource assignment (dictated to the drivers directly by an ideal centralized scheme) and conditions are derived for minimizing the related price of
anarchy. Finally, the numerical results and the presented discussion provide hints
for the practical management and pricing of public and private parking resources.

## 1 Introduction

The tremendous increase of urbanization necessitates the efficient and environmentally
sustainable management of urban processes and operations. The (pilot) implementation of solutions and pioneering ideas from the area of information and communication
technologies presents a response to this need paving the way for the so-called "smart
cities". The search for available parking space is among the daily routine processes
that can benefit from this new kind of city environments. In particular, transportation
engineers have developed parking assistance systems, realized through *information dissemination mechanisms* to alleviate not only the traffic congestion problems that stem
from the blind parking search but also the resulting environmental burden. Common to

---

* This work has been supported in part by the European Commission IST-FET project RECOG-
NITION (FP7-IST-257756), the European Commission under the Network of Excellence in
Internet Science project EINS (FP7-ICT-288021) and the University of Athens Special Ac-
count of Research Grants no 10812.

W. Elmenreich, F. Dressler, and V. Loreto (Eds.): IWSOS 2013, LNCS 8221, pp. 54–65, 2014.

these systems is the exploitation of wireless communications and information sensing technologies to collect and broadcast (in centralized mechanisms, *i.e.*, in [18], [20]) or share (in distributed systems, *i.e.*, in [8], [10]) information about the availability of parking space and the demand for it within the search area. This information is then used to steer the parking choices of drivers in order to reduce the effective competition over the parking space and make the overall search process efficient. Additionally, the implementation of smart demand-responsive pricing policies on the parking facilities intends to improve parking availability in overused parking zones and reduce double-parking and cruising phenomena (*i.e.*, in [1]).

This paper seeks to systematically explore the fundamental limits on the efficiency of these parking assistance systems. To this end, it ideally assumes that drivers become *completely* aware of the competition intensity, parking capacity and applied pricing policies on the parking facilities. Thus, the particular questions that become relevant for this exploration are: How do these three parameters modulate drivers' parking choices? How do they affect the cost that drivers incur and the revenue accruing for the parking service operator?

We formulate the parking spot selection problem as an instance of *resource selection games* in Section 2. We view the drivers as rational selfish agents that try to minimize the cost they pay for acquired parking space. The drivers choose to either compete for the cheaper but scarce on-street parking spots or head for the more expensive private parking lot(s)[1]. In the first case, they run the risk of failing to get a spot and having to *a posteriori* take the more expensive alternative, this time suffering the additional *cruising* cost in terms of time, fuel consumption (and stress) of the failured attempt. Drivers make their decisions drawing on perfect information about the number of drivers, the availability of parking spots and the pricing policy, which is broadcast from the parking service operator. In Section 3, we derive the equilibrium behaviors of the drivers and compare the induced social cost against the optimal one via the Price of Anarchy metric. Most importantly, in Section 4, we show that the optimization of the equilibrium social cost is feasible by properly choosing the pricing and location of the private parking facilities. We outline related research in Section 5 and we close the discussion in Section 6 iterating on the model assumptions.

## 2   The Parking Spot Selection Game

In the parking spot selection game, the set of players consists of drivers who circulate within the center area of a big city in search of parking space. Typically, in these regions, parking is completely forbidden or constrained in whole areas of road blocks so that the real effective curbside is significantly limited (see Fig. 1). The drivers have to decide whether to drive towards the scarce low-cost (controlled) public parking spots or the over-dimensioned and more expensive private parking lot (we see all local lots collectively as one). All parking spots that lie in the same public or private area are assumed to be of the same value for the players. Especially the on-street parking spots

---

[1] Hereafter in the paper, the terms *public* parking spots and *private* parking facilities denote *on-street* parking spots and parking lots, respectively. Their context in this paper should not be confused with that of public/private goods in economics.

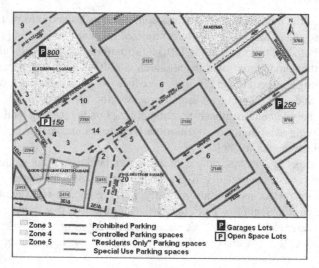

**Fig. 1.** Parking map of the centre area of Athens, Greece. Dashed lines show metered controlled (*public*) parking spots, whereas "P" denotes *private* parking facilities. The map illustrates, as well, the capacity of both parking options.

are considered quite close to eachother, resulting in practically similar driving times to them and walking times from them to the drivers' ultimate destinations. Thus, the decisions are made on the two *sets* of parking spots rather than individual set items.

We observe drivers' behavior within a particular time window over which they reach this parking area. In general, these synchronization phenomena in drivers' flow occur at specific time zones during the day [2]. Herein, we account for morning hours or driving in the area for business purposes coupled with long parking duration. Thus, the collective decision making on parking space selection can be formulated as an instance of the strategic *resource selection games*, whereby $N$ players (*i.e.*, drivers) compete against each other for a finite number of common resources (*i.e.*, public parking spots) [6]. More formally, the one-shot[2] parking spot selection game is defined as follows:

**Definition 1.** *A* Parking Spot Selection Game *is a tuple* $\Gamma = (\mathcal{N}, \mathcal{R}, (A_i)_{i \in \mathcal{N}}, (w_j)_{j \in (pub, priv)})$, *where:*

- $\mathcal{N} = \{1, ..., N\}, N > 1$ *is the set of drivers who seek for parking space,*
- $\mathcal{R} = \mathcal{R}_{pub} \cup \mathcal{R}_{priv}$ *is the set of parking spots;* $\mathcal{R}_{pub}$ *is the set of public spots, with* $R = |\mathcal{R}_{pub}| \geq 1$; $\mathcal{R}_{priv}$ *the set of private spots, with* $|\mathcal{R}_{priv}| \geq N$,
- $A_i = \{public, private\}$, *is the action set for each driver* $i \in \mathcal{N}$,
- $w_{pub}(\cdot)$ *and* $w_{priv}(\cdot)$ *are the cost functions of the two actions, respectively.*

The parking spot selection game comes under the broader family of *congestion games*. The players' payoffs (here: costs) are non-decreasing functions of the *number* of players competing for the parking capacity rather than their identities and common

---

[2] The study of the dynamic variant of the game is a natural direction for future work.

to all players. More specifically, drivers who decide to compete for the public parking space undergo the risk of not being among the $R$ winner-drivers to get a spot. In this case, they have to eventually resort to private parking space, only after wasting extra time and fuel (plus patience supply) on the failed attempt. The expected cost of the action $public$, $w_{pub} : A_1 \times ... \times A_N \to \mathbb{R}$, is therefore a function of the number of drivers $k$ taking it, and is given by

$$w_{pub}(k) = min(1, R/k)c_{pub,s} + (1 - min(1, R/k))c_{pub,f} \tag{1}$$

where $c_{pub,s}$ is the cost of successfully competing for public parking space, whereas $c_{pub,f} = \gamma \cdot c_{pub,s}, \gamma > 1$, is the cost of competing, failing, and eventually paying for private parking space.

On the other hand, the cost of private parking space is fixed

$$w_{priv}(k) = c_{priv} = \beta \cdot c_{pub,s} \tag{2}$$

where $1 < \beta < \gamma$, so that the excess cost $\delta \cdot c_{pub,s}$, with $\delta = \gamma - \beta > 0$, reflects the actual cost of cruising and the "virtual" cost of wasted time till eventually heading to the private parking space.

We denote an action profile by the vector $a = (a_i, a_{-i}) \in \times_{k=1}^{N} A_k$, where $a_{-i}$ denotes the actions of all other drivers but player $i$ in the profile $a$. Besides the two *pure* strategies reflecting the pursuit of public and private parking space, the drivers may also randomize over them. In particular, if $\Delta(A_i)$ is the set of probability distributions over the action set of player $i$, a player's *mixed action* corresponds to a vector $p = (p_{pub}, p_{priv}) \in \Delta(A_i)$, where $p_{pub}$ and $p_{priv}$ are the probabilities of the pure actions, with $p_{pub} + p_{priv} = 1$, while its cost is a weighted sum of the cost functions $w_{pub}(\cdot)$ and $w_{priv}(\cdot)$ of the pure actions.

In the following section, we derive the game-theoretic analysis of the particular game formulation looking into both the stable and optimal operational conditions as well as the respective costs incurred by the players.

## 3   Game Analysis

Ideally, the players determine their strategy under complete knowledge of those parameters that shape their cost. Given the symmetry of the game, the additional piece of information that is considered available to the players, besides the number of vacant parking spots and the employed pricing policy, is the level of parking demand, *i.e.*, the number of drivers searching for parking space. We draw on concepts from [15] and theoretical results from [6, 9] to derive the equilibrium strategies for the game $\Gamma$ and assess their (in)efficiency.

### 3.1   Pure Equilibrium Strategies

*Existence:* The parking spot selection game constitutes a symmetric game, where the action set is common to all players and consists of two possible actions, *public* and *private*. Cheng *et al.* have shown in ([9], Theorem 1) that every symmetric game with two strategies has an equilibrium in pure strategies.

*Derivation:* Due to the game's symmetry, the full set of $2^N$ different action profiles maps into $N + 1$ different action meta-profiles. Each meta-profile $a(m), m \in [0, N]$ encompasses all $\binom{N}{m}$ different action profiles that result in the same number of drivers competing for on-street parking space. The expected costs for these $m$ drivers and for the $N - m$ ones choosing directly the private parking lot alternative are functions of $a(m)$ rather than the exact action profile.

In general, the cost $c_i^N(a_i, a_{-i})$ for driver $i$ under the action profile $a = (a_i, a_{-i})$ is

$$c_i^N(a_i, a_{-i}) = \begin{cases} w_{pub}(N_{pub}(a)), & \text{for } a_i = public \\ w_{priv}(N - N_{pub}(a)), & \text{for } a_i = private \end{cases} \tag{3}$$

where $N_{pub}(a)$ is the number of competing drivers for on-street parking under action profile $a$. Equilibrium action profiles combine the players' *best-responses* to their opponents' actions. Formally, an action profile $a = (a_i, a_{-i})$ is a pure Nash equilibrium if for all $i \in \mathcal{N}$:

$$a_i \in \arg\min_{a_i' \in A_i} (c_i^N(a_i', a_{-i})) \tag{4}$$

so that no player has anything to gain by changing her decision unilaterally.

Therefore, to derive the equilibrium states, we determine the conditions on $N_{pub}$ that break the equilibrium definition and reverse them. More specifically, given an action profile $a$ with $N_{pub}(a)$ competing drivers, a player gains by changing her decision to play action $a_i$ in two circumstances:

$$\text{when } a_i = private \text{ and } w_{pub}(N_{pub}(a) + 1) < c_{priv} \tag{5}$$
$$\text{when } a_i = public \text{ and } w_{pub}(N_{pub}(a)) > c_{priv} \tag{6}$$

Taking into account the relation between the number of drivers and the available on-street parking spots, $R$, we can postulate the following Lemma:

**Lemma 1.** *In the parking spot selection game $\Gamma$, a driver is motivated to change her action $a_i$ in the following circumstances:*

- $a_i = private$ *and* $N_{pub}(a) < R \leq N$ *or* $\tag{7}$

$$R \leq N_{pub}(a) < N_0 - 1 \leq N \text{ or} \tag{8}$$

$$N_{pub}(a) < N \leq R \tag{9}$$

- $a_i = public$ *and* $R < N_0 < N_{pub}(a) \leq N$ $\tag{10}$

*where* $N_0 = \frac{R(\gamma - 1)}{\delta} \in \mathbb{R}$.

*Proof.* Conditions (7) and (9) are trivial. Since the current number of competing vehicles is less than the on-street parking capacity, every driver having originally chosen the private parking option has the incentive to change her decision due to the price differential between $c_{pub,s}$ and $c_{priv}$.

When $N_{pub}(a)$ exceeds the public parking supply, a driver who has decided to avoid competition, profits from switching her action when (5) holds, which combined with (1), yields (8). Similarly, a driver that first decides to compete, profits by switching her action if (6) holds, which combined with (1), yields (10).

The following Theorem for the pure Nash equilibria of the parking spot selection game may now be stated.

**Theorem 1.** *A parking spot selection game has:*

(a) *one Nash equilibrium* $a^*$ *with* $N_{pub}(a^*) = N_{pub}^{NE,1} = N$, *if* $N \leq N_0$ *and* $N_0 \in \mathbb{R}$

(b) $\binom{N}{\lfloor N_0 \rfloor}$ *Nash equilibrium profiles* $a'$ *with* $N_{pub}(a') = N_{pub}^{NE,2} = \lfloor N_0 \rfloor$, *if* $N > N_0$ *and* $N_0 \in (R, N) \backslash \mathbb{N}^*$

(c) $\binom{N}{N_0}$ *Nash equilibrium profiles* $a'$ *with* $N_{pub}(a') = N_{pub}^{NE,2} = N_0$ *and* $\binom{N}{N_0-1}$ *Nash equilibrium profiles* $a^*$ *with* $N_{pub}(a^*) = N_{pub}^{NE,3} = N_0 - 1$, *if* $N > N_0$ *and* $N_0 \in [R+1, N] \cap \mathbb{N}^*$.

*Proof.* Theorem 1 follows directly from (4) and Lemma 1. The game has two equilibrium conditions on $N_{pub}$ for $N > N_0$ with integer $N_0$ (case c), or a unique equilibrium condition, otherwise (cases a, b). ∎

*Efficiency:* The efficiency of the equilibrium action profiles resulting from the strategically selfish decisions of the drivers is assessed through the broadly used metric of the Price of Anarchy [15]. It expresses the ratio of the social cost in the worst-case equilibria over the optimal social cost under ideal coordination of the drivers' strategies.

**Proposition 1.** *In the parking spot selection game, the* pure *Price of Anarchy equals:*

$$
PoA = \begin{cases} \dfrac{\gamma N - (\gamma-1)\min(N,R)}{\min(N,R)+\beta\max(0,N-R)}, & \text{if } N_0 \geq N \\[3mm] \dfrac{\lfloor N_0 \rfloor \delta - R(\gamma-1)+\beta N}{R+\beta(N-R)}, & \text{if } N_0 < N \end{cases}
$$

*Proof.* The social cost under action profile $a$ equals:

$$
C(N_{pub}(a)) = \sum_{i=1}^{N} c_i^N(a) =
$$

$$
c_{pub,s}(N\beta - N_{pub}(a)(\beta - 1)), \text{ if } N_{pub}(a) \leq R \text{ and} \quad (11)
$$

$$
c_{pub,s}(N_{pub}(a)\delta - R(\gamma - 1) + \beta N), \text{ if } R < N_{pub}(a) \leq N
$$

The numerators of the two ratios are obtained directly by replacing the first two $N_{pub}^{NE}$ values (a) and (b) (worst-cases) computed in Theorem 1. On the other hand, under the socially optimal action profile $a_{opt}$, exactly $R$ drivers pursue on-street parking, whereas the remaining $N - R$ go directly for the private parking. Therefore, under $a_{opt}$, no drivers find themselves in the unfortunate situation to have to pay the additional cost of cruising in terms of time and fuel after having unsuccessfully competed for an on-street parking spot. The optimal social cost, $C_{opt}$ is given by:

$$
C_{opt} = \sum_{i=1}^{N} c_i^N(a_{opt}) = c_{pub,s}[min(N, R) + \beta \cdot max(0, N - R)]
$$

**Proposition 2.** *In the parking spot selection game, the* pure *Price of Anarchy is upper-bounded by* $\frac{1}{1-R/N}$ *with* $N > R$.

*Proof.* The condition is obtained directly from Proposition 1, when $N > R$. ∎

## 3.2  Mixed-Action Equilibrium Strategies

We mainly draw our attention on *symmetric* mixed-action equilibria since these can be more helpful in dictating practical strategies in real systems. Asymmetric mixed-action equilibria are discussed in [14].

*Existence:* Ashlagi, Monderer, and Tennenholtz proved in ([6], Theorem 1) that a unique symmetric mixed equilibrium exists for the broader family of resource selection games with more than two players and increasing cost functions. It is trivial to repeat their proof and confirm this result for our parking spot selection game $\Gamma$, with $N > R$ and cost functions $w_{pub}(\cdot)$ and $w_{priv}(\cdot)$ that are non-decreasing functions of the number of players (increasing and constant, respectively).

*Derivation:* Let $p = (p_{pub}, p_{priv})$ denote a mixed-action. Then, the expected costs of choosing the on-street (resp. private) parking space option, when all other drivers play the mixed-action $p$, are given by

$$c_i^N(public, p) = \sum_{N_{pub}=0}^{N-1} w_{pub}(N_{pub} + 1)B(N_{pub}; N - 1, p_{pub}) \qquad (12)$$

$$c_i^N(private, p) = c_{priv} \qquad (13)$$

where $B(N_{pub}; N-1, p_{pub})$ is the Binomial probability distribution with parameters $N - 1$ and $p_{pub}$, for $N_{pub}$ drivers choosing on-street parking. The cost of the symmetric profile where everyone plays the mixed-action $p$ is given by

$$c_i^N(p, p) = p_{pub} \cdot c_i^N(public, p) + p_{priv} \cdot c_i^N(private, p) \qquad (14)$$

With these at hand, we can now postulate the following Theorem.

**Theorem 2.** *The parking spot selection game $\Gamma$ has a unique symmetric mixed-action Nash equilibrium $p^{NE} = (p_{pub}^{NE}, p_{priv}^{NE})$, where:*

- $p_{pub}^{NE} = 1$, *if $N \leq N_0$ and*
- $p_{pub}^{NE} = \frac{N_0}{N}$, *if $N > N_0$,*

*where $p_{pub}^{NE} = 1 - p_{priv}^{NE}$ and $N_0 \in \mathbb{R}$.*

*Proof.* The symmetric equilibrium for $N \leq N_0$ corresponds to the pure NE we derived in Theorem 1. To compute the equilibrium for $N > N_0$ we invoke the condition that equilibrium profiles must fulfil

$$c_i^N(public, p^{NE}) = c_i^N(private, p^{NE}) \qquad (15)$$

namely, the costs of each pure action belonging to the support of the equilibrium mixed-action strategy are equal. Hence, from (12), (13) and (15) the symmetric mixed-action equilibrium $p^{NE} = (p_{pub}^{NE}, p_{priv}^{NE})$ solves the equation

$$f(p) = -\beta + \sum_{k=0}^{N-1} (\gamma - \min(1, \frac{R}{k+1}) \cdot (\gamma - 1))B(k; N - 1, p) = 0 \qquad (16)$$

A closed-form expression for the equilibrium $p_{pub}^{NE}$ is not straightforward. However, it holds that:

$$\lim_{p \to 0} f(p) = -\beta + 1 < 0 \text{ and } \lim_{p \to 1} f(p) = \delta(1 - \frac{N_0}{N}) > 0 \qquad (17)$$

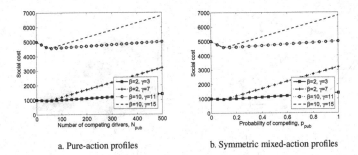

a. Pure-action profiles                b. Symmetric mixed-action profiles

**Fig. 2.** Social cost for $N = 500$ drivers when exactly $N_{pub}$ drivers compete (a) or when all drivers decide to compete with probability $p_{pub}$ (b), for $R = 50$ public parking spots, under different charging policies

and $f(p)$ is a continuous and strictly increasing function in $p$ since

$$f'(p) = \sum_{k=0}^{N-1} (\gamma - \min(1, \frac{R}{k+1})(\gamma - 1))B'(k; N-1, p) > \sum_{k=0}^{N-1} B'(k; N-1, p) = 0$$

Hence, $f(p)$ has a single solution. It may be checked with replacement that $f(N_0/N) = 0$.

## 4   Numerical Results

The analysis in Section 3 suggests that the charging policy for on-street and private parking space and their relative location, which determines the overhead parameter $\delta$ of failed attempts for on-street parking space, affect to a large extent the (in)efficiency of the game equilibrium profiles. In the following, we illustrate their impact on the game outcome and discuss their implications for real systems.

For the numerical results we adopt per-time unit normalized values used in the typical municipal parking systems in big European cities [2]. The parking fee for public space is set to $c_{pub,s} = 1$ unit whereas the cost of private parking space $\beta$ ranges in $(1, 16]$ units and the excess cost $\delta$ in $[1, 5]$ units. We consider various parking demand levels assuming that private parking facilities in the area suffice to fulfil all parking requests.

Figure 2 plots the social costs $C(N_{pub})$ under pure (Eq. 11) and $C(p_{pub})$ under mixed-action strategies as a function of the number of competing drivers $N_{pub}$ and competition probability $p_{pub}$, respectively, where

$$C(p_{pub}) = c_{pub,s} \sum_{n=0}^{N} \binom{N}{n} p_{pub}^n (1 - p_{pub})^{N-n} \cdot$$
$$[min(n, R) + max(0, n - R)\gamma + (N - n)\beta] \qquad (18)$$

Figure 2 motivates two remarks. First, the social cost curves for pure and mixed-action profiles have the same shape. This comes as no surprise since for given $N$, any value for

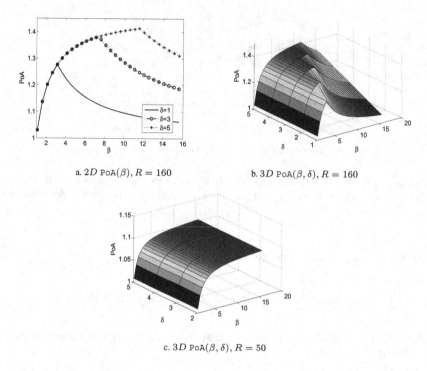

a. $2D$ PoA$(\beta)$, $R = 160$

b. $3D$ PoA$(\beta, \delta)$, $R = 160$

c. $3D$ PoA$(\beta, \delta)$, $R = 50$

**Fig. 3.** Price of Anarchy for $N = 500$ and varying $R$, under different charging policies

the expected number of competing players $0 \leq N_{pub} \leq N$ can be realized through appropriate choice of the symmetric mixed-action profile $p$. Second, the cost is minimized when the number of competing drivers is equal to the number of on-street parking spots. The cost rises when either competition exceeds the available on-street parking capacity or drivers are overconservative in competing for on-street parking. In both cases, the drivers pay the penalty of the lack of coordination in their decisions. The deviation from optimal grows faster with increasing price differential between the on-street and private parking space.

Whereas an optimal centralized mechanism would assign exactly $\min(N, R)$ public parking spots to $\min(N, R)$ drivers, if $N > R$, in the worst-case equilibrium the size of drivers' population that actually competes for on-street parking spots exceeds the real parking capacity by a factor $N_0$ which is a function of $R$, $\beta$ and $\gamma$ (equivalently, $\delta$) (see Lemma 1). This inefficiency is captured in the PoA plots in Figure 3 for $\beta$ and $\delta$ ranging in $[1.1, 16]$ and $[1, 5]$, respectively. The plots illustrate the following trends:

**Fixed $\delta$ - varying $\beta$:** For $N \leq N_0$ or, equivalently, for $\beta \geq \frac{\delta(N-R)+R}{R}$, it holds that $\frac{\vartheta PoA}{\vartheta \beta} < 0$ and therefore, the PoA is strictly decreasing in $\beta$. On the contrary, for $\beta < \frac{\delta(N-R)+R}{R}$, the PoA is strictly increasing in $\beta$, since $\frac{\vartheta PoA}{\vartheta \beta} > 0$.

**Fixed $\beta$ - varying $\delta$:** For $N \leq N_0$ or, equivalently, for $\delta \leq \frac{R(\beta-1)}{N-R}$ we get $\frac{\vartheta PoA}{\vartheta \delta} >$ 0. Therefore, the PoA is strictly increasing in $\delta$. For $\delta > \frac{R(\beta-1)}{N-R}$, we get $\frac{\vartheta PoA}{\vartheta \delta} = 0$. Hence, if $\delta$ exceeds $\frac{R(\beta-1)}{N-R}$, PoA is insensitive to changes of the excess cost $\delta$.

Practically, the equilibrium strategy emerging from this kind of assisted parking search behavior, approximates the optimal coordinated mechanism when the operation of private parking facilities accounts for drivers' preferences as well as estimates of the typical parking demand and supply. More specifically, if, as part of the pricing policy, the cost of private parking is less than $\frac{\delta(N-R)+R}{R}$ times the cost of on-street parking, then the social cost in the equilibrium profile approximates the optimal social cost as the price differential between public and private parking decreases. This result is inline with the statement in [16], arguing that *"price differentials between on-street and off-street parking should be reduced in order to reduce traffic congestion"*. Note that the PoA metric also decreases monotonically for high values of the private parking cost, specifically when the private parking operator desires to gain more than $\frac{\delta(N-R)+R}{R}$ times the cost of on-street parking, towards a bound that depends on the excess cost $\delta$. Nevertheless, these operating points correspond to high absolute social cost, *i.e.*, the minimum achievable social cost is already unfavorable due to the high fee paid by $N - R$ drivers that use the private parking space (see Fig. 2, large $\beta$). On the other hand, there are instances, as in case of $R = 50$ (see Fig. 3), where the value $\frac{\delta(N-R)+R}{R}$ corresponds to a non-realistic option for the cost of private parking space, already for $\delta > 1$. Thus, contrary to the previous case, PoA only improves as the cost for private parking decreases. Finally, for given cost of the private parking space, the social cost can be optimized by locating the private facility in the proximity of the on-street parking spots so that the additional travel distance is reduced to the point of bringing the excess cost below $\frac{R(\beta-1)}{N-R}$.

## 5   Related Work

Various aspects of the broader parking space search problem have been addressed in the literature. Probably the first formulation of the problem appeared in [17] in the context of the broader family of stopping problems. Parking spots are spread randomly, with density $\lambda$ over equal-size blocks that are adjacent to a destination, and drivers circle through them crossing the destination every time such a circle is over. Ferguson in [11] considers a simpler variant of the problem, whereby the drivers' destination lies in the middle of an infinite-length straight line with parking spots that are occupied with probability $p$. In either case, the optimal policy for the drivers is shown to be of the threshold type: they should occupy an available vacant parking spot whenever this lies within some distance $r = f(\lambda)$, resp. $f(p)$, from their destination and continue searching otherwise.

Pricing and more general economic dimensions of the parking allocation problem are analyzed in [16] and [3]. A queueing model for drivers who circulate in search for on-street parking is introduced in [16] in order to analyze economic effects of congestion pricing. From a microeconomical point of view, Anderson and de Palma in [3] treat the parking spots as common property resource and question whether free access or some pricing structure result in more efficient use of the parking capacity. Working on a simple model of city and parking spot distribution, they show that this use is more efficient (in fact, optimal) when the spots are charged with the fee chosen in the monopolistically competitive equilibrium with private ownership. The situation is reversed, *i.e.*, drivers are better off when access to the parking spots is free of charge.

Subsequent research contributions in [5], [4] and [7] have explicitly catered for strategic behavior and the related game-theoretic dimensions of general parking applications. In [5], the games are played among parking facility providers and concern the location and capacity of their parking facility as well as which pricing structure to adopt. Whereas, in the other two works, the strategic players are the drivers. In [4], which seeks to provide cues for optimal parking lot size dimensioning, the drivers decide on the arriving time at the lot, accounting for their preferred time as well as their desire to secure a space. In a work more relevant to ours, the authors in [7] define a game setting where drivers exploit (or not) information on the location of others to occupy an available parking spot at the minimum possible travelled distance, irrespective of the distance between the spot and driver's actual travel destination. The authors present distributed parking spot assignment algorithms to realize or approximate the Nash equilibrium states. In our work, the game-theoretic analysis highlights the cost effects of the parking operators' pricing policies on drivers' decisions, drawing on *closed-form expressions* of the stable operational points in the different game settings.

## 6    Conclusions - Discussion

In this paper, we draw our attention on fundamental determinants of parking assistance systems' efficiency rather than particular realizations. We have, thus, formulated the information-assisted parking search process as an instance of resource selection games to assess the ultimate impact that an ideal mechanism broadcasting accurate information on parking demand, supply and pricing policies, can have on drivers' decisions. Our model dictates plausible conditions under which different pricing policies steer the equilibrium strategies, reduce the inefficiency of the parking search process, and favor the social welfare.

We conclude by iterating on the strong and long-debatable assumption that drivers *do* behave as fully rational utility-maximizing decision-makers; namely, they can exhaustively analyze the possible strategies available to themselves and the other drivers, identify the equilibrium profile(s), and act accordingly to realize it. Simon in [19], challenged both the *normative* and *descriptive* capacity of the fully rational decision-maker, arguing that human decisions are most often made under knowledge, time and computational constraints. One way to accommodate the first constraints is through *(pre-) Bayesian* games of incomplete information. In [14], we formulate (pre-)Bayesian variants of the parking spot selection game to assess the impact of information accuracy on drivers' behavior and, ultimately, the service cost. However, models that depart from the utility-maximization norm and draw on fairly simple *cognitive heuristics*, e.g., [12], reflect better Simon's argument that humans are satisficers rather than maximizers. For example, the authors in [13] explore the impact of the *fixed-distance* heuristic on a simpler version of the unassisted parking search problem. The comparison of normative and more descriptive decision-making modeling approaches both in the context of the parking spot selection problem and more general decision-making contexts, is an interesting area worth of further exploration.

# References

[1] http://sfpark.org/
[2] http://www.city-parking-in-europe.eu/
[3] Anderson, S.P., de Palma, A.: The economics of pricing parking. Journal of Urban Economics 55(1), 1–20 (2004)
[4] Arbatskaya, M., Mukhopadhaya, K., Rasmusen, E.: The parking lot problem. Tech. rep., Department of Economics, Emory University, Atlanta (2007)
[5] Arnott, R.: Spatial competition between parking garages and downtown parking policy. Transport Policy (Elsevier), 458–469 (2006)
[6] Ashlagi, I., Monderer, D., Tennenholtz, M.: Resource selection games with unknown number of players. In: Proc. AAMAS 2006, Hakodate, Japan (2006)
[7] Ayala, D., Wolfson, O., Xu, B., Dasgupta, B., Lin, J.: Parking slot assignment games. In: Proc. 19th ACM SIGSPATIAL GIS (2011)
[8] Caliskan, M., Graupner, D., Mauve, M.: Decentralized discovery of free parking places. In: Proc. 3rd VANET (in Conjunction with ACM MobiCom), Los Angeles, CA, USA (2006)
[9] Cheng, S.G., Reeves, D.M., Vorobeychik, Y., Wellman, M.P.: Notes on the equilibria in symmetric games. In: Proc. 6th Workshop on Game Theoretic and Decision Theoretic Agents (Colocated with IEEE AAMAS), New York, USA (August 2004)
[10] Delot, T., Cenerario, N., Ilarri, S., Lecomte, S.: A cooperative reservation protocol for parking spaces in vehicular ad hoc networks. In: Proc. 6th Inter. Conference on Mobile Technology, Application and Systems, MOBILITY (2009)
[11] Ferguson, T.S.: Optimal stopping and applications, http://www.math.ucla.edu/~tom/Stopping/
[12] Goldstein, D.G., Gigerenzer, G.: Models of ecological rationality: The recognition heuristic. Psychological Review 109(1), 75–90 (2002)
[13] Hutchinson, J.M.C., Fanselow, C., Todd, P.M.: Ecological rationality: intelligence in the world. Oxford University Press, New York (2012)
[14] Kokolaki, E., Karaliopoulos, M., Stavrakakis, I.: Leveraging information in vehicular parking games. Tech. rep., Dept. of Informatics and Telecommunications, Univ. of Athens (2012), http://cgi.di.uoa.gr/~evako/tr1.pdf
[15] Koutsoupias, E., Papadimitriou, C.H.: Worst-case equilibria. Computer Science Review 3(2), 65–69 (2009)
[16] Larson, R.C., Sasanuma, K.: Congestion pricing: A parking queue model. Journal of Industrial and Systems Engineering 4(1), 1–17 (2010)
[17] MacQueen, J., Miller, J.: Optimal persistence policies. Operations Research 8(3), 362–380 (1960)
[18] Mathur, S., Jin, T., Kasturirangan, N., Chandrasekaran, J., Xue, W., Gruteser, M., Trappe, W.: Parknet: Drive-by sensing of road-side parking statistics. In: Proc. 8th ACM MobiSys., San Francisco, CA, USA, pp. 123–136 (2010)
[19] Simon, H.A.: A behavioral model of rational choice. The Quarterly Journal of Economics 69(1), 99–118 (1955)
[20] Wang, H., He, W.: A reservation-based smart parking system. In: Proc. 5th International Workshop on Cyber-Physical Networking Systems (Colocated with IEEE INFOCOM) (April 2011)

# The Relative Disagreement Model of Opinion Dynamics: Where Do Extremists Come From?

Michael Meadows and Dave Cliff

Department of Computer Science, University of Bristol, Bristol BS8 1UB, U.K.
michaeljmeadows86@gmail.com, dc@cs.bris.ac.uk

**Abstract.** In this paper we introduce a novel model that can account for the spread of extreme opinions in a human population as a purely local, self-organising process. Our starting point is the well-known and influential Relative Agreement (RA) model of opinion dynamics introduced by Deffuant *et al.* (2002). The RA model explores the dynamics of opinions in populations that are initially seeded with some number of "extremist" individuals, who hold opinions at the far ends of a continuous spectrum of opinions represented in the abstract RA model as a real value in the range [-1.0, +1.0]; but where the majority of the individuals in the population are, at the outset, "moderates", holding opinions closer to the central mid-range value of 0.0. Various researchers have demonstrated that the RA model generates opinion dynamics in which the influence of the extremists on the moderates leads, over time, to the distribution of opinion values in the population converging to attractor states that can be qualitatively characterised as one of either uni-polar and bi-polar extremes, or reversion to the centre ("central convergence"). However, a major weakness of the RA model is that it pre-supposes the existence of extremist individuals, and hence says nothing to answer the question of "where do extremists come from?" In this paper, we introduce the Relative Disagreement (RD) model, in which extremist individual arise spontaneously and can then exert influence over moderates, forming large groups of polar extremists, via an entirely internal, self-organisation process. We demonstrate that the RD model can readily exhibit the uni-polar, bi-polar, and central-convergence attractors that characterise the dynamics of the RA model, and hence this is the first paper to describe an opinion dynamic model in which extremist positions can spontaneously arise and spread in a population via a self-organising process where opinion-influencing interactions between any two individuals are characterised not only by the extent to which they agree, but also by the extent to which they disagree.

**IWSOS Key Topics:** Models of self-organisation in society, techniques and tools for modelling self-organising systems, self-organisation in complex networks, self-organising group and pattern formation, social aspects of self-organisation.

## 1 Introduction: Opinion Dynamics and the RA Model

Since its inception, "opinion dynamics" has come to refer to a broad class of different models applicable to many fields ranging from sociological phenomena to ethology and physics (Lorenz 2007). The focus of this paper is on an improvement of Deffuant

W. Elmenreich, F. Dressler, and V. Loreto (Eds.): IWSOS 2013, LNCS 8221, pp. 66–77, 2014.

*et al.*'s (2002) "Relative Agreement" (RA) model, that was originally developed to assess the dynamics of political, religious and ideological extremist opinions, and the circumstances under which those opinions can rise to dominance via processes of self-organisation (i.e., purely by local interactions among members of a population) rather than via exogenous influence (i.e. where the opinion of each member of a population is influenced directly by an external factor, such as mass-media propaganda). The RA model was developed with the aim of helping to explain and understand the growth of extremism in human populations, an issue of particular societal relevance in recent decades where extremists of various religious or political beliefs have been linked with significant terrorist acts.

Suppose a group of *n* experts are tasked with reaching an agreement on a given subject. Initially, all the experts will possess an opinion that for simplicity we imagine can be represented as a real number *x*, marking a point on some continuum. During the course of their meeting, the experts present their opinion to the group in turn and then modify their own opinion in light of the views of the others, by some fixed weight. If all opinions are equal after the interaction, it can be said that a consensus has been reached, otherwise another round is required. It was demonstrated by de Groot (1974) that this simple model would always reach a consensus for any positive weight. Although highly abstract and clearly not particularly realistic, this simple model has become the basis for further analysis and subsequent models (e.g. Chatterjee & Seneta 1977; Friedkin 1999).

Building on the de Groot model, the Hegselmann-Krause model included the additional constraint that the experts would only consider the opinions of others that were not too dissimilar from their own (Krause 2000); this is also known as the Bounded Confidence (BC) model. The BC model adds the idea that each expert has a quantifiable conviction about their opinion, their uncertainty, *u*. It was demonstrated that although a consensus may be reached in the BC model, it is not guaranteed (Hegselmann & Krause 2002). It was observed that when the BC model is set in motion with every agent having an initially high confidence (low uncertainty) about their own random opinion, the population disaggregates into large numbers of small clusters; and as the uncertainty was increased, so the dynamics of the model tended towards those of the original de Groot model (Krause 2000). Later, the model was tested with the inclusion of "extremist" agents, defined as individuals having extreme value opinions and very low uncertainties. In the presence of extremists it was found that the population could tend towards two main outcomes: *central convergence* and *bipolar convergence* (Hegselmann & Krause 2002). In central convergence, typical when uncertainties are low, the majority of the population clustered around the central, "moderate" opinion. In contrast, when uncertainties were initially high, the moderate population would split into two approximately equal groups one of which would tend towards the positive extreme and the other towards the negative: referred to as *bipolar convergence*.

Although these two phenomena have occurred in real human societies, there is a third well-known phenomenon that the BC model is unable to exhibit: an initially moderate population tending towards a single extreme (and hence known as *single extreme convergence*).

Shortly after the publication of the BC model, Deffuant, Amblard, Weisbuch, & Faure (2002) reported their exploration of the BC model and proposed an extension of it which they named the Relative Agreement (RA) model (Deffuant *et al.* 2002). The RA model was intended to be capable of exhibiting single extreme convergence in its dynamics.

There are two main differences between the RA model and the BC model. The first change is that agents are no longer expressing their opinion to the group as a whole followed by a group-wide opinion update. Instead, in the RA model pairs of agents are randomly chosen to interact and update. This is repeated until stable clusters have formed. The second change relates to how agents update their opinions. In the BC model an agent only accepted an opinion if it fell within the bounds of their own uncertainty, and the weight that was applied to that opinion was fixed. In the RA model however, an opinion is weighted proportional to the degree of overlap between the uncertainties of the two interacting agents.

These changes represent a push for increased realism. In large populations, individuals cannot necessarily consider the opinion of every other agent; therefore paired interactions are far more plausible. More importantly, the RA model also allows for agents with strong convictions to be far more convincing than those who are uncertain (Deffuant 2006). Thus, although the RA model is stochastic, the only random element of the model is in the selection of the individuals for the paired interactions (Lorenz 2005). As expected, the RA model was able to almost completely replicate the key results of the BC model (Deffuant *et al.* 2000).

Having demonstrated that RA model was comparable to the BC model under normal circumsthances, Deffuant *et al.* then added the *extremist* agents to the population, to see if they could cause shifts in the opinions of entire population. An extremist was defined as an agent with an opinion above 0.8 or below -0.8 and with a very low uncertainty. Conversely, a moderate agent is one whose absolute opinion value is less than 0.8 and with a fixed, higher uncertainty who is therefore more willing to be persuaded by other agents. Under these circumstances, Deffuant *et al.* reported that there are large areas of parameter space in which all three main types of population convergence could occur. The fact that the RA model offers realistic parameter-settings under which single extreme convergence regularly occurs is a particularly novel attraction.

To classify population convergences, Deffuant *et al.* (2002) introduced the $y$ metric, defined as: $y = {p'_+}^2 + {p'_-}^2$ where $p'_+$ and $p'_-$ are the proportion of initially moderate agents that have finished with an opinion that is classified as extreme at the positive and negative end of the scale respectively. Thus, central, bipolar and single extreme convergences have $y$ values of 0.0, 0.5 and 1.0, respectively.

Meadows & Cliff (2012) recently demonstrated that while all three population convergences can indeed occur using various parameter settings, the specific parameter values that do allow for the more extreme convergences are not as originally reported by Deffuant *et al.* (2002). Meadows & Cliff state that they reviewed over 150 papers that cite Deffuant *et al.* (2002) but not a single one of them involved an independent replication of the original published RA results. In attempting to then replicate the results reported by Deffuant *et al.* (2002), Meadows & Cliff (2012) found a significantly different, but simpler and more intuitively appealing result, as is illustrated here in Figures 1 and 2. In Figure 2, as agents' initial uncertainty $u$ increases, the instability of the population rises with it, resulting in a higher $y$ value; also, as the proportion of initially extremist agents increases, there is again a corresponding rise in the resultant instability. When there is a higher level of instability in the population, Figure 2 shows that there is a greater chance of the population converging in a manner other than simply to the centre. Also, in Figure 2 (unlike Figure 1) there is no area that implies a *guaranteed* single extreme convergence, although there is a large area of parameter space in which that can occur. This makes intuitive sense because a population with an initial instability that would allow single extreme convergence must surely also be unstable enough to allow central and bipolar convergences.

## 2    The Relative Disagreement Model

While the ability to produce a single extreme is significant, it should be noted that the RA model requires a significant minority (20-30%) of extremists be seeded into the initial population. Although useful as an academic tool, it could be argued that this is a somewhat artificial solution as it simply raises further questions, most notably, "Where did all of those extremist agents come from in the first place?" A model that could exhibit similar results to the RA model without the need for such a high proportion of extremist agents would be seen as a major improvement. One that could do away with the extremist agents altogether would be a considerable leap forward. The remainder of this paper will focus on the specification and examination of a model, quite like the RA model, that achieves similar results without the need for initially extreme agents.

The crux of this improvement lies in the observation that the RA model focuses only on the behaviour of agents when they are in agreement, yet it has long been known to psychologists that disagreements in opinions can lead to the distancing of the two opinions. That is to say, if you are debating with someone whose opinion is completely different to your own, you are likely to "update" your own opinion even further away from that with which you disagree. This is called *reactance* (Baron & Byrne 1991) and can be thought of as analogous to the more general "Boomerang effect" (Brehm & Brehm 1981) where someone attempting to influence an individual inadvertently produces the opposite effect in attitude change. Given this additional real-world information, there follows a formal definition of the proposed "relative disagreement" (RD) model; after the formal definition, we present an empirical exploration of its opinion dynamics.

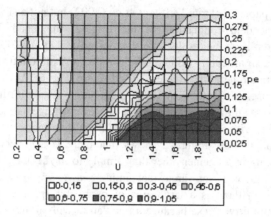

**Fig. 1.** Average $y$ values over the $(p_e, u)$ parameter-space, reproduced directly from Deffuant *et al.* (2002), Figure 9

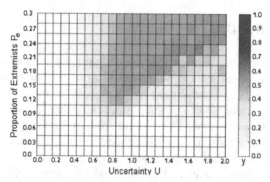

**Fig. 2.** Average $y$ values over the $(p_e, u)$ parameter-space, as reported by Meadows & Cliff (2012); the zone of parameter space, and the values of other RA parameters, make this plot a direct correlate of the original shown in Figure 1. Reproduced directly from Meadows & Cliff (2012) Figure 12.

If we return to our population of $n$ agents, each individual $i$ is in possession of two variables; an opinion $x$, and an uncertainty $u$, both of which are represented by real numbers. In the RA model, the opinion could be initially set on the range of -1.0 to 1.0, with extremists being defined as agents whose opinions lay below -0.8 or above 0.8. As our goal is to replicate the convergences on the RA model without extremist agents we do not allow an opinion to be initially set outside the range of -0.8 and 0.8, but we retain the minimal and maximal values as before. Since we have no extremist agents, we are no longer constrained by defining agents by their opinions and so uncertainties are assigned randomly using a simple method to bias agents towards being uncertain (as it is in uncertain populations that more interesting results are to be found) given by:

$$u = min( random(0.2, 2.0) + random(0.0, 1.0), 2.0)$$

Random paired interactions take place between agents until a stable opinion state is produced. The relative agreement between agents $i$ and $j$ is calculated as before by taking the overlap between the two agents' bounds $h_{ij}$, given by:

$$h_{ij} = min\ (x_i + u_i, x_j + u_j) - max(x_i - u_i, x_j - u_j)$$

Followed by subtracting the size of the non-overlapping part given by:

$$2u_i - h_{ij}$$

So the total agreement between the two agents is given as:

$$h_{ij} - (2u_i - h_{ij}) = 2(h_{ij} - u_i)$$

Once that is calculated, the relative agreement is then given by:

$$2(h_{ij} - u_i)\ /\ 2u_i = (h_{ij} / u_i) - 1$$

Then if $h_{ij} > u_i$, then update of $x_j$ and $u_j$ is given by:

$$x_j := x_j + \mu_{RA}[(h_{ij} / u_i) - 1](x_i - x_j)$$
$$u_j := u_j + \mu_{RA}[(h_{ij} / u_i) - 1](u_i - u_j)$$

Similarly, the relative disagreement between agents $i$ and $j$ is calculated by a very similar method to find $g_{ij}$:

$$g_{ij} = max(x_i - u_i, x_j - u_j) - min\ (x_i + u_i, x_j + u_j)$$

Followed by subtracting the size of the non-overlapping part given by:

$$2u_i - g_{ij}$$

So the total disagreement between the two agents is given as:

$$g_{ij} - (2u_i - g_{ij}) = 2(g_{ij} - u_i)$$

Once that is calculated, the relative disagreement is then given by:

$$2(g_{ij} - u_i)\ /\ 2u_i = (g_{ij} / u_i) - 1$$

We have chosen to use an analogous method for calculating the agents' disagreement for ease of understanding as it also facilitates the need for calculating relative disagreement. Now we would not want the disagreement update to occur in every instance of disagreement, as it is intuitively obvious that this would not occur in the every real-world instance of disagreement. Therefore if $g_{ij} > u_i$ and with a probability $\lambda$, the update of $x_j$ and $u_j$ is given by:

$$x_j := x_j - \mu_{RD}[(g_{ij} / u_i) - 1](x_i - x_j)$$
$$u_j := u_j + \mu_{RD}[(g_{ij} / u_i) - 1](u_i - u_j)$$

Note that with this model, the definition of $g_{ij}$ will be equal to that of $h_{ij}$, but it is important to express this definition in full for a greater intuitive understanding of the model and the logic behind its construction.

# 3    Results of the RD Model

## 3.1    Comparing the RA and RD Model

As has been stated previously, the RD model can only be considered to be an improvement of the RA model if it can replicate the results of the RA model without the aid of initially extreme agents. This comparison however, leads to an initial problem that is worth noting now. When analysing the RA model we can examine individual sample runs (as shown in the next section) or typical patterns of $y$ like those in Figure 1 and 2.

As shown those figures, we see that the main parameters examined when we look for various patterns of $y$ are the initial uncertainty of the moderate agents (equivalent to all agents in the RD model) and the proportion of initially extreme agents in the population, which clearly presents us with a problem. Neither of those values are relevant to the RD model; uncertainties are randomised and there are no longer any extremist agents. Therefore the patterns of $y$ that we shall find with the RD model will not be analogous with those from the RA model. Indeed, in terms of comparing the RA and RD models, the first and most convincing step will be to demonstrate that all three convergences are possible before we move on to looking at new methods of examining $y$-values.

## 3.2    Reproducing Convergences

Clearly the most basic function of any model aiming to be comparable to the RA model is to be able to demonstrate the three main population opinion convergences: central, bipolar and single extreme. In addition, a successful model should be able to perform these convergences without any drastic alterations to the parameters in the original RA model, much like the specification provided in Section 2. As would be expected, we find that in this model a population converges towards the centre in a majority of the simulation runs, more so than in the RA model. We believe that this is understandable because we no longer have 30% of the agents functioning solely to skew the overall opinion of the population towards their own. Any aberration in the population's opinion must come from agreements and importantly, disagreements from within the body of moderate agents. Nonetheless, we found that central and bipolar convergences can occur regularly in given parameter spaces and that a single extreme convergence can occur when the uncertainties are randomised as prescribed previously as shown in Figures 3, 4, and 5.

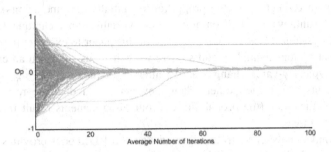

**Fig. 3.** An example of central convergence in the RD model with n = 200, U = 0.8, $\mu_{RA}$ = 0.05, $\mu_{RD}$ = 0.05, $\lambda$ = 0. 0

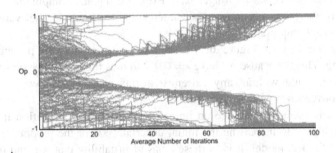

**Fig. 4.** An example of bipolar convergence in the RD model with n = 200, U = 0.4, $\mu_{RA}$ = 0.2, $\mu_{RD}$ = 0.2, $\lambda$ = 0.1

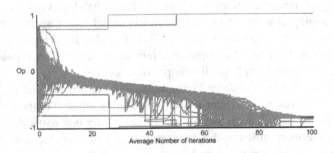

**Fig. 5.** An example of single extreme convergence in the RD model with n = 200, U = 0.1, $\mu_{RA}$ = 0.5, $\mu_{RD}$ = 0.7, $\lambda$ = 0.4

## 3.3    Examining New Patterns of *y*

Having already established that the heat-map graphs visualising how the $y$-value changes with respect to $p_e$ (the proportion of initially extreme agents) and $u$ are no longer relevant, we need to employ a new visualisation technique for illustrating patterns of $y$-values. It is beyond the scope of this paper to examine the whole variety of different options available, and so here we will focus solely on an examination of how the typical $y$-value changes as we alter the values $\mu_{RA}$ and $\mu_{RD}$ (the degree to which agents affect each other when they agree and disagree respectively) with different values of $\lambda$ (the probability that our disagreements result in an action), as shown in Figure 6.

It is immediately clear from Figure 6 that the RD model provides much higher population stability than the RA model. Note that almost every $y$-value heat-map is almost entirely white, indicating a very high proportion of simulation runs that resulted in central convergence. This is of course not a disappointing result and should be expected; we no longer have extremist agents manipulating our moderate agents. It therefore becomes more interesting to examine the various standard deviation heat-map graphs.

In the top row of Figure 6, we see that there is nothing particularly special happening. This is because we have $\lambda = 0.0$, and thus the model behaves exactly as the RA model would without any extremist agents and as such, we see a consistent central convergence with no variation.

Once $\lambda$ is increased, we see an immediate increase in population instability; and effect analogous to increasing the initial uncertainty $u$, or the proportion of extremist agents, in the RA model. It is in these areas of instability that we find the possibility of a population converging in a bipolar or single extreme manner although, as with the RA model, these instabilities do not guarantee that bipolar or single extreme convergences *will* occur, merely that they *could* occur. That is: there is an increased probability of such instabilities occurring but the probability is not unity. Not only is this encouraging in terms of validating the RD model with respect to the RA model findings given by Meadows & Cliff (2012), but also in terms of realism: in the majority of model runs, the population remain moderate with bipolar and single extreme populations being a small minority of outcomes.

A final point of interest from Figure 6 would be to observe that the increase in $\lambda$ significantly increases the size and level of instability inherent to the population. This is of course to be expected, because as we see in any one particular graph, as $\mu_{RD}$ increases, the more unstable the population becomes, implying that increasing the influence that a disagreement can have will increase the population. It also explains the observation that as $\lambda$ increases, the need for $\mu_{RD}$ to be a high value to allow for population instability decreases quite substantially.

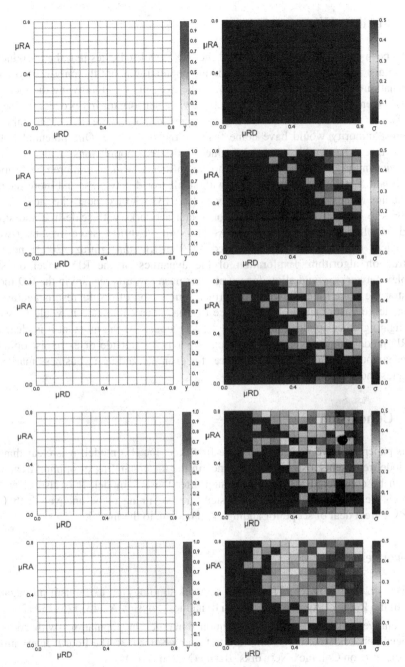

**Fig. 6.** Average $y$ values (left column) and standard deviations (right column) for (from top to bottom) $\lambda = 0.0, 0.25, 0.5, 0.75,$ and $1.0$

## 4     Further Work

As has been shown, this examination of the RD model indicates that the introduction of extremist agents in an otherwise moderate population is not the only way to create a population-wide instability. By removing the extremists and allowing the moderate agents to generate the instability themselves, via a self-organising process, we see an increase in realism while answering the question of where the significant RA extremist minority would have come from in the first place. One potential line of further research would be to introduce clustered populations with nontrivial topologies in the "social network" of who can influence who, as explored by Amblard & Deffuant (2004) who experimented with small-world networks; and more recently by [Author names deleted for blind review] (2013) who extended the study of RA networks to look at opinion dynamics on social networks with scale-free and small-world topologies, and also on networks with "hybrid" scale-free/small-world topological characteristics as generated by the Klemm-Eguiluz (2002) network construction algorithm. Exploration of the dynamics of the RD model on such complex networks would allow for yet another push towards realism in the RD model whilst at the same time allowing for further comparison of results results against the RA model to ensure applicability.  Once that has been completed it would be worth investigating other application areas for the RD model. At the moment it is clear that the RD model could be applied to much larger fields outside of extremist opinions and behaviour relating to terrorist nature (for example, aiding in business marketing strategies and techniques).

## 5     Conclusion

In this paper we have presented, for the first time, the RD model. It is clear that the RD model represents a significant advance in the understanding of opinion dynamics and can be considered a logical next step after the RA model. It is also clear that although the properties of this new system are beginning to be uncovered, there is already a great deal of scope for adding extra realism to the model.

## References

Amblard, F., Deffuant, G.: The role of network topology on extremism propagation with the Relative Agreement opinion dynamics. Physica A 343, 725–738 (2004)

The Relative Agreement model of opinion dynamics in populations with complex social network structure. Submiited to CompleNet 2013: The Fourth International Workshop on Complex Networks, Berlin (March 2013)

Baron, R., Byrne, D.: Social Psychology, 6th edn. Allyn and Bacon, Boston (1991)

Brehm, S., Brehm, J.: Psychological reactance: a theory of freedom and control. Academic Press, New York (1981)

Chatterjee, S., Seneta, E.: Towards Consensus: Some Convergence Theorems on Repeated Averaging. Journal of Applied Probability 14(1), 88–97 (1977)

De Groot, M.: Reaching a Consensus. Journal of the American Statistical Association 69(345), 118–121 (1974)

Deffuant, G., Neau, D., Amblard, F.: Mixing beliefs among interacting agents. Advances in Complex Systems 3, 87–98 (2000); Deffuant, G., Amblard, F., Weisbuch, G., Faure, T.: How can extremism prevail? A study based on the relative agreement interaction model. Journal of Artificial Societies and Social Simulation 5(4), 1 (2002)

Deffuant, G.: Comparing Extremism Propagation Patterns in Continuous Opinion Models. Journal of Artificial Societies and Social Simulation 9(3), 8 (2006)

Friedkin, N.: Choice Shift and Group Polarization. American Sociological Review 64(6), 856–875 (1999)

Hegselmann, R., Krause, U.: Opinion dynamics and bounded confidence: models, analysis and simulation. Journal of Artificial Societies and Social Simulation 5(3), 2 (2002)

Klemm, K., Eguíluz, V.: Growing Scale-Free Networks with Small-World Behavior. Physical Review 65, 057102 (2002)

Kozma, B., Barrat, A.: Consensus formation on adaptive networks. Physical Review E 77, 016102 (2008)

Krause, U.: A Discrete Nonlinear and Non-Autonomous Model of Consensus Formation. In: Communications in Difference Equations: Proceedings of the Fourth International Conference on Difference Equations, August 27-31, 1998, pp. 227–236 (2000)

Lorenz, J., Deffuant, G.: The role of network topology on extremism propagation with the relative agreement opinion dynamics. Physica A: Statistical Mechanics and Its Applications 343, 725–738 (2005)

Lorenz, J.: Continuous Opinion Dynamics under Bounded Confidence: A Survey. International Journal of Modern Physics C 18, 1–20 (2007)

Meadows, M., Cliff, D.: Reexamining the Relative Agreement Model of Opinion Dynamics. Journal of Artificial Societies and Social Simulation 15(4), 4 (2012)

Sobkowicz, P.: Studies on opinion stability for small dynamic networks with opportunistic agents. International Journal of Modern Physics C 20(10), 1645–1662 (2009)

# Modeling the Emergence of a New Language: Naming Game with Hybridization

Lorenzo Pucci[1], Pietro Gravino[2,3], and Vito D.P. Servedio[2]

[1] Phys. Dept., Univ. Federico II, Complesso Monte S. Angelo, 80126 Napoli, Italy
[2] Phys. Dept., Sapienza Univ. of Rome, P.le A. Moro 2, 00185 Roma, Italy
[3] Phys. Dept., Alma Mater Studiorum, Univ. of Bologna, Italy

**Abstract.** In recent times, the research field of language dynamics has focused on the investigation of language evolution, dividing the work in three evolutive steps, according to the level of complexity: lexicon, categories and grammar. The Naming Game is a simple model capable of accounting for the emergence of a lexicon, intended as the set of words through which objects are named. We introduce a stochastic modification of the Naming Game model with the aim of characterizing the emergence of a new language as the result of the interaction of agents. We fix the initial phase by splitting the population in two sets speaking either language A or B. Whenever the result of the interaction of two individuals results in an agent able to speak both A and B, we introduce a finite probability that this state turns into a new idiom C, so to mimic a sort of hybridization process. We study the system in the space of parameters defining the interaction, and show that the proposed model displays a rich variety of behaviours, despite the simple mean field topology of interactions.

## 1 Emergence of a Lexicon as a Language

The modeling activity of language dynamics aims at describing language evolution as the global effect of the local interactions between individuals in a population of $N$ agents, who tend to align their verbal behavior locally, by a negotiation process through which a successful communication is achieved [1, 2]. In this framework, the emergence of a particular communication system is not due to an external coordination, or a common psychological background, but it simply occurs as a convergence effect in the dynamical processes that start from an initial condition with no existing words (agents having to invent them), or with no agreement.

Our work is based on the Naming Game (NG) model, and on its assumptions [3]. In Fig. 1 we recall the NG basic pairwise interaction scheme. A fundamental assumption of NG is that vocabulary evolution associated to every single object is considered independent. This lets us simplify the evolution of the whole lexicon as the evolution of the set of words associated to a single object, equally perceived in the sensorial sphere by all agents.

The simplicity of NG in describing the emergence of a lexicon also relies on the fact that competing words in an individual vocabulary are not weighted, so that

W. Elmenreich, F. Dressler, and V. Loreto (Eds.): IWSOS 2013, LNCS 8221, pp. 78–89, 2014.

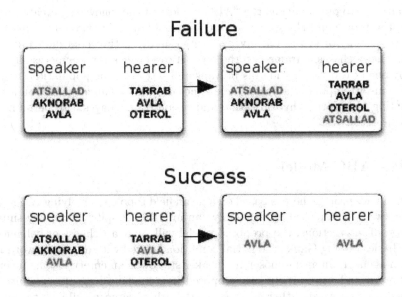

**Fig. 1. Naming Game interaction scheme.** A *speaker* and a *hearer* are picked up randomly. The speaker utters a word chosen randomly in his vocabulary. *Failure*: the hearer does not know the uttered word and he simply adds it to his vocabulary. *Success*: the hearer knows the uttered word and both agents by agreement delete all the words in their vocabularies except the uttered word one.

they can be easily stored or deleted. In fact it turns out that for convergence to a single word consensus state, the weights are not necessary as it was supposed by the seminal work in this research field [3]. Every agent is a box that could potentially contain an infinite number of words, so the number of states that can define the agent is limited only by the number of words diffused in the population at the beginning of the process (so that anyone can speak with at least one word).

In this work we aimed not only at the aspect of competition of a language form with other ones, but also at introducing interactions between them, with the possibility of producing new ones. We investigate conditions for the success of a new idiom, as product of a synthesis and competition of preexisting idioms. To this purpose, we introduce a stochastic interaction scheme in the basic Naming Game accounting for this synthesis, and show in a very simple case that the success of the new spoken form at expense of the old ones depends both on the stochastic parameters and the fractions of the different idioms spoken by populations at the beginning of the process. We have simulated this process starting from an initial condition where a fraction $n_A$ of the population speaks with $A$ and the remaining $n_B$ with $B$. It turns out that when the different-speaking fractions are roughly of the same size the new form, which we call $C$ (therefore we shall refer to our model as the "ABC model" in the following), created from the synthesis of $A$ and $B$, establishes and supplants the other two forms. Instead, when $n_A > n_B$ (or symmetrically $n_B > n_A$), above a threshold depending on the chosen stochastic parameters, the term $A$ establishes (or symmetrically $B$), namely one of the starting idioms prevails and settles in the population.

Previous attempts to model the emergence of a Creole language, i.e. an idiom originating by a sort of hybridization and fusion of languages, can be found in literature [4–6].

## 2   The ABC Model

The model we propose here is based on a mean field topology involving $N$ agents, i.e. any two agents picked up randomly can interact. Despite this pretty simple topology of interactions, the proposed model will show a richness of behaviors. In the basic Naming Game, the initial condition is fixed with an empty word list for each agent. If an agent chosen as speaker still owns an empty list, he invents a new word and utters it. In our proposed new model all agents are initially assigned a given word, either $A$ or $B$, so that there is no possibility to invent a brand new word, unless an agent with both $A$ and $B$ in his list is selected. In that case, we introduce a finite probability $\gamma$ that his list containing $A$ and $B$ turns into a new entry $C$ (Fig. 2). We interpret $\gamma$ as a measure of the need of agents to obtain a sort of hybridization through which they understand each other, or a measure of the natural tendency of two different language forms to be synthesized together. In the latter case, different values of $\gamma$ would depend on the language forms considered in a real multi language scenario.

The stochastic rule $AB \rightarrow C$ may be applied either in the initial interaction phase by changing the dictionary of the speaker, or at the final stage by changing the state of the hearer after the interaction with the speaker. In this paper we show the results obtained by running the trasformation $AB \rightarrow C$ (and also $ABC \rightarrow C$ when a speaker has got an $ABC$ vocabulary) before the interaction. Introducing the trasformation after the interaction changes the results from a qualitative point of view, producing only a shift of transition lines between the final states in the space of parameters defining the stochastic process.

A further stochastic modification of the basic Naming Game interaction, firstly introduced in [7], has also been adopted here. It gives account for the emergence or persistence of a multilingual final state, where more than one word is associated to a single object. This is done by mimicking a sort of confidence degree among agents: in case of a successful interaction, namely when the hearer shares the word uttered by the speaker, the update trasformation of the two involved vocabularies takes place with probability $\beta$ (the case $\beta = 1$ obviously reduces to the basic NG). Baronchelli et al. [7] showed in their model (which corresponds to our model at $\gamma = 0$) that a transition occurs around $\beta = 1/3$. For $\beta > 1/3$ the system converges to the usual one word consensus state (in this case only $A$ or $B$). For $\beta < 1/3$ the system converges to a mixed state, where more than one word remains, so that there exist single and multi word vocabularies at the end of the

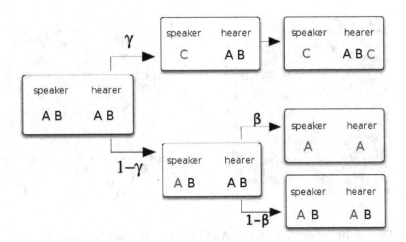

**Fig. 2. ABC model:** Example of interaction with the introduction of the trasformation of $AB \to C$ with probability $\gamma$ on the speaker vocabulary. The trasformation is
carried before the usual speaker-hearer NG interaction, which includes the stochastic
update of the vocabularies in case of a success depending on $\beta$.

process (namely $A$, $B$ and $AB$). A linear stability analysis of the mean field master equations of this model (describing the evolution of vocabulary frequencies
$n_A$, $n_B$ and $n_{AB}$) shows that the steady state $n_A = 1$, $n_B = n_{AB} = 0$ (or symmetrically $n_B = 1$, $n_A = n_{AB} = 0$), which is stable for $\beta > 1/3$, turns unstable
if $\beta < 1/3$, where viceversa the steady state $n_A = n_B = b(\beta)$, $n_{AB} = 1 - 2b(\beta)$
emerges as a stable state, with $b(\beta)$ being a simple algebraic expression of the
parameter $\beta$. In our work, as shown next, we found that the transition order-
disorder (i.e. single word vs. multi word final state) at $\beta = 1/3$ remains for all
the values of $\gamma$.

The numerical stochastic simulation of the process for selected values of
$(\beta, \gamma) \in [0, 1] \times [0, 1]$, indicates that the system presents a varied final phase
space as shown in Fig. 3 and 4, left panel. The transition line at $\beta = 1/3$ remains: for $\beta > 1/3$ the system converges to a one-word final state, with a single
word among $A$, $B$ and $C$, while for $\beta < 1/3$ it converges to a multi-word state
with one or more than one word spoken by each agent. This result is confirmed
by the integration of the mean field master equation of the model, describing
the temporal evolution of the fractions of all vocabulary species $n_A$, $n_B$, $n_C$,
$n_{AB}$, $n_{AC}$, $n_{BC}$, $n_{ABC}$, present in the system. The results of such integration,
involving a fourth order Runge-Kutta algorithm, display the same convergence
behaviour of the stochastic model (Fig. 3 and 4, right panel), though they are
obviously characterized by less noise.

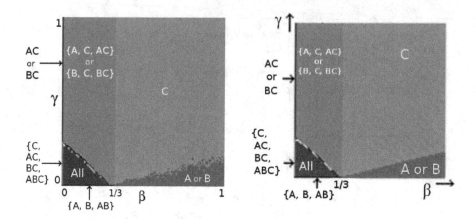

**Fig. 3. Phase diagram $\gamma, \beta$ of the ABC model:** The new language $C$ is created at the beginning of the interaction by updating the speaker's vocabulary. We have $0 \leq \beta \leq 1$ and $0 \leq \gamma \leq 1$ on the horizontal and vertical axis, respectively. *Left:* Results of the stochastic model where the number of agents involved is $N = 1000$ and the final time is set to $10^6$. The initial condition on populations is fixed as $n_A \simeq n_B$ and $n_{AB} = 0$ at $t = 0$. *Right:* Results of the mean field master equation numerically integrated till a final time $t = 1000$. The initial condition where chosen as $n_A = 0.49$ and $n_B = 0.51$ (and symmetrically $n_A = 0.51$ and $n_B = 0.49$). We employed a fourth order Runge-Kutta algorithm with temporal step $dt = 0.1$.

## 2.1   High confidence $\beta > 1/3$

By looking at Fig. 3 at the region with $\beta > 1/3$, we note a transition interval between the final state composed only of either $A$ or $B$ (red region) and a final state with only $C$ (orange region). The fuzziness of the border dividing these two domains, evident in the left panel of the figure, can be ascribed to finite size effects, for the separation line gets sharper by enlarging the number of agents $N$, eventually collapsing towards the strict line obtained by the integration of the mean field master equation (right panel of the figure), which we report in the Appendix section. The linear stability analysis of the cumbersome mean field master equation reveals that these two phases are both locally stable, and turn unstable when $\beta < 1/3$. The convergence to one or the other phase depends on the initial conditions, i.e. whether the system enters the respective actraction basins during its dynamical evolution. To demonstrate this, we studied the behaviour of the system by fixing $\beta$ and varying both $\gamma$ and the initial conditions $n_A = \alpha$, $n_B = 1 - \alpha$, with $\alpha \in [0, 1]$. The result reported in Fig. 4 clearly shows a dependence on the initial conditions. In particular, if the initial condition $|n_A - n_B| = 1 - 2\alpha$ is sufficiently large, the convergence to the $C$ phase disappears. The corresponding threshold value of $\alpha$ decreases slightly by decreasing the value of parameter $\beta$, going from $\alpha = 0.34$ for $\beta = 1$ to $\alpha = 0.24$ for $\beta = 0.34$ (i.e. slightly above the transition signalled by $\beta = 1/3$).

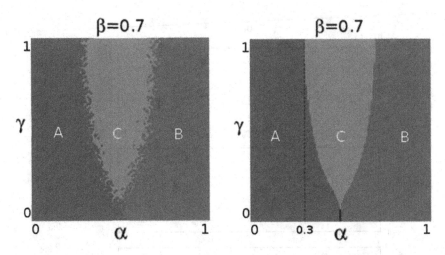

**Fig. 4. Phase diagram $\gamma$, $\alpha$ of the ABC model:** Final phase diagram depending on $\gamma$ and initial condition $n_A = 1 - \alpha$ and $n_B = \alpha$ with $\alpha \in [0, 1]$, with $\beta = 0.7$ fixed. *Left:* diagram obtained after $10^6$ pair interactions of the stochastic process. *Right:* diagram obtained by integrating up to $t = 1000$ the mean field master equation with a fourth order Runge-Kutta algorithm employing a temporal step $dt = 0.1$. For $\alpha < 0.3$ (or symmetrically $\alpha > 0.7$) the system converges to the $n_A = 1$ (or symmetrically $n_B = 1$) for every considered value of $\gamma$.

By numerically solving the mean field master equation, we analyzed the evolution of the system in proximity of the transition between the (orange) region characterized by the convergence to the $C$ state and the (red) region where the convergence is towards either the state $A$ or $B$, being these latter states discriminated by the initial conditions $n_A > 1/2$ or $n_A < 1/2$ respectively. We show the result in Fig. 5 obtained by fixing $\beta = 0.8$ and $\alpha = 0.49$. Initially, both the fractions of $n_A$ (black curve) and $n_B$ (red curve) decrease in favour of $n_{AB}$ (blue curve). Thereafter, also $n_{AB}$ starts to decrease since the mixed $AB$ and $ABC$ states can be turned directly into $C$, causing an increase of $n_C$ (green curve). While the $n_{AB}$, $n_{ABC}$, $n_A$, $n_{AC}$ fractions vanish quite soon, mainly because fewer agents have the $A$ state in their vocabulary, the states involving $B$ and $C$ survive, reaching a meta-stable situation in which $n_C = n_B \approx 0.37$ and $n_{BC} \approx 0.26$. This meta-stable state of the system is clearly visible in the mid panel of Fig. 5. The life time of the meta-stable state diverges by approaching the corresponding set of parameters ($\gamma = \gamma_c \approx 0.12234742$ in the Figure; $\gamma_c$ depends on $\beta$), with the result that the overall convergence time diverges as well. The stochastic simulation would differ from the solution of the master equation right at this point: a small random fluctuation in the fraction of either $B$ or $C$ will cause the convergence of the stochastic model towards one or the other state, while the deterministic integration algorithm (disregarding computer numerical errors) will result always in the win of the state $C$ for $\gamma > \gamma_c$ or the state $B$ for $\gamma < \gamma_c$. The fuzziness visible in all the figures related to the stochastic model is a direct consequence of those random fluctuations.

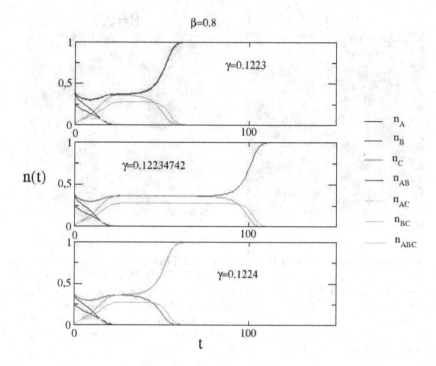

**Fig. 5. Temporal evolution of the ABC model:** Temporal evolution of all vocabulary species occurring in the system calculated by integrating the mean field master equation. The initial population was chosen as $n_B = 0.51$ and $n_A = 0.49$, while we fixed $\beta = 0.8$. For $\gamma = 0.1223$ the system converges to the $B$ consensus state, while for $\gamma = 0.1224$ the system converges to the $C$ consensus state. The initial trivial transient with the formation of the $AB$ mixed states has been removed. By approaching the transition point the convergence time increases.

Another interesting area in the $\gamma, \beta$ phase space of the model is the boundary between the region around $\beta = 1/3$, where one switches from a convergence to a single state ($\beta > 1/3$) to a situation with the coexistence of multiple states ($\beta < 1/3$). As $\beta \to 1/3^+$ the time of convergence towards the $C$ consensus phase, which is the absorbing state whenever $\gamma > 0$, diverges following a power-law with the same exponent as in the case of $\gamma = 0$, where we recover the results of [7], i.e. $t_{\text{conv}} \simeq (\beta - 1/3)^{-1}$ (Fig. 6). Of course, in the case $\gamma = 0$ there is no $C$ state involved anymore and the competition is only between the $A$ and $B$ states. Moreover, as we note from Fig. 6, the convergence time to the one word consensus state is the highest when $\gamma = 0$ and decreases by increasing the value of $\gamma$. This result is somewhat counter intuitive since we expect that the presence of three states $A$, $B$, $C$ would slow down the convergence with respect to a situation with only two states $A$ and $B$, but actually in the first case another supplementary channel is yielding the stable $C$ case, i.e. the $AB \longrightarrow C$ channel (neglecting of course the rare $ABC \longrightarrow C$) thus accelerating the convergence.

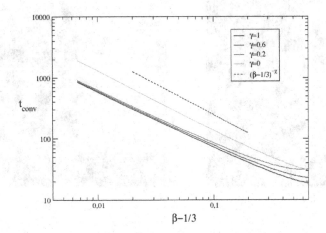

**Fig. 6. Convergence time in the ABC model for** $\beta \to 1/3^+$**:** Time of convergence of the system to the one word consensus state as obtained with the mean field master equation as a function of $(\beta - 1/3)$. The initial population densities were set to $n_A = 0.49$ and $n_B = 0.51$. Importantly, if $\gamma > 0$ the system converges to the hybrid state $C$, while for $\gamma = 0$ the state $B$ prevails (orange curve). The dotted line refers to a power-law behaviour of exponent $-1$. We used a fourth order Runge-Kutta algorithm with temporal step $dt = 0.1$.

The linearization of the mean field master equation around the absorbing points with $\beta > 1/3$ delivers six negative eigenvalues, confirming that the points in the orange and red region of Fig. 3 are locally stable. Moreover, it comes out that those eigenvalues do not depend on $\gamma$ showing that the choice of the initial conditions on $n_A$ and $n_B$ is crucial in entering the two different actraction basins. As a consequence of this independence on $\gamma$, the equation of the line dividing the orange and red regions cannot be calculated easily.

## 2.2   Low Confidence $\beta < 1/3$

In the case $\beta < 1/3$ we get multi-word final states. The green color in Fig. 3 stands for an asymptotic situation where $n_B = n_C$ and $n_{BC} = 1 - n_B - n_C$ (and of course a symmetric situation with $B$ replaced by $A$ when the initial conditions favour $A$ rather than $B$). The dependence of the asymptotic fractions $n_B$ and $n_C$ on $\beta$ is the same of that occurring for $\gamma = 0$ and presented in [7].

Instead, the blue color of Fig. 3 represents an asymptotic state where all vocabulary typologies are present $A$, $B$, $C$, $AB$, $BC$, $AC$ and $ABC$, with $n_A = n_B$ and $n_{AC} = n_{BC}$. In this case the vocabulary fractions depend both on $\beta$ and $\gamma$. White dots in the left panel of Fig. 3, which tend to disappear enlarging the population size $N$, are points where the system has not shown a clear stable state

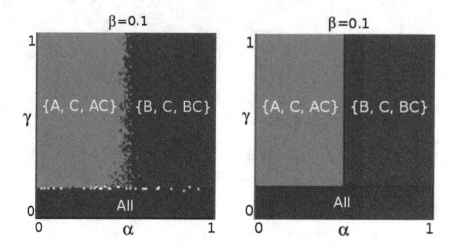

**Fig. 7. Phase diagram $\gamma, \alpha$ of the ABC model:** Phase diagram depending on $\gamma$ and initial condition $n_A = 1 - \alpha$ and $n_B = \alpha$ with $\alpha \in [0, 1]$, with $\beta = 0.1$ fixed. *Left:* diagram obtained after $t = 10^6$ steps of the stochastic process. *Right:* diagram obtained by integrating up to $t = 1000$ the mean field master equation with a fourth order Runge-Kutta algorithm employing a temporal step $dt = 0.1$.

after the chosen simulation time. In fact, they disappear in the final phase space described by the mean field master equation. Contrary to the case of $\beta > 1/3$, the final state does not depend on the particular initial conditions provided that initially $n_A + n_B = 1$. By fixing $\beta = 0.1$ and varing $\gamma$ and initial conditions $\alpha \in [0, 1]$, we get the steady behavior shown in Fig. 7.

The linearization of the mean field master equation around the absorbing points with $\beta < 1/3$ in the green region of Fig. 3 reveals that those are attractive points (six negative eigenvalues) irrespective of the initial condition provided that $n_A + n_B = 1$. The equation of the transition line that divides the blue and green region can be inferred numerically, again with the linearization of the master equation. In particular the transition point at $\beta = 0$ can be found analytically to be at $\gamma = 1/4$. The independence on the initial conditions makes the region $\beta < 1/3$ substantially different from the complementary region $\beta > 1/3$.

## 3   Conclusions

We modeled the emergence of a new language $C$ as a result of the mutual interaction of two different populations of agents initially speaking different idioms $A$ and $B$ and interacting each other without restriction (mean field). Such tight connections between individuals speaking two different idioms is certainly unrealistic, but the same reasoning can be extended to accomplish the birth of single hybrid words resulting from the interaction of two pronunciation variants of the same object (eg. the english word *rowel*, perhaps an hybridization of the latin *rota* and the english *wheel*).

Three parameters govern the time evolution of the model and characterize the final asymptotic state: $\beta$ the measure of the tendency of a hearer to adopt the shared word used by the speaker (confidence), $\gamma$ the probability that two forms $A$ and $B$ are synthesized into the form $C$, and $\alpha$ the initial condition in the space $n_A + n_B = 1$. It turns out that:

- for $\beta < 1/3$ the system converges to multi-word states, all containing a fraction of the state $C$, and that do not depend on the initial conditions provided that $n_A + n_B = 1$.
- for $\beta > 1/3$ the system converges to a consensus state where all agents end up with the same state, either $A$, $B$ or $C$. The transition line $\gamma(\beta)$ separating the $A$ or $B$ convergence state from $C$, which are all locally stable independently from $\gamma$, depends on the initial distribution of $n_A$ and $n_B$, with $n_A + n_B = 1$. Moreover, the invention of $C$ produces a reduction of the time of convergence to the consensus state (all agents speaking with $C$) when starting with an equal fraction of $A$ and $B$ in the population.

Interestingly the modern point of view of linguists links the birth and continuous development of all languages as product of local interaction between the varied language pools of individuals who continuously give rise to processes of competition and exchange between different forms, but also creation of new forms in order to get an arrangement with the other speakers [8]. In this view of a language as mainly a social product it seems that the use of the Naming Game is particularly fit, in spite of the old conception of pure languages as product of an innate psychological background of individuals [9].

It would be interesting to apply our model in the study of real language phenomena where a sort of hybridization of two or more languages in a contact language ecology takes ground. There are many examples of this in the history, as for example the formation of the modern romance European languages from the contact of local Celtic populations with the colonial language of Romans, Latin. A more recent example of this is the emergence of Creole languages in colonial regions where European colonialists and African slaves came into contact [10].

The starting point for a comparison of our model with this kind of phenomena would be retrieving demographic data of the different ethnic groups at the moment they joined in the same territory and observing if a new language established. Our point of view would be obviously not to understand how particular speaking forms emerged, but to understand whether there is a correlation between the success of the new language forms and the initial language demography. In this case, a more refined modeling would take into account also the temporal evolution of the population due to reproduction and displacements, and the particular topologies related to the effective interaction scheme acting in the population.

**Acknowledgements.** The authors wish to thank V. Loreto and X. Castelló for useful discussions. The present work is partly supported by the EveryAware european project grant nr. 265432 under FP7-ICT-2009-C.

## Appendix: Mean Field Master Equation

The mean field master equation of the ABC model, in the case in which the speaker changes her vocabulary with the rule $\{AB, ABC\} \xrightarrow{\gamma} \{C\}$ before the interaction, is the following:

$$
\begin{aligned}
\frac{dn_A}{dt} =\ & -n_A n_B - n_C n_A + \left((1-\gamma)\frac{\beta-1}{2}+\beta-\gamma\right) n_{AB} n_A + \frac{3\beta-1}{2} n_{AC} n_A - n_{BC} n_A + \\
& + (1-\gamma)\beta n_{AB}^2 + \beta n_{AC}^2 + \beta\left((1-\gamma)+1\right) n_{AB} n_{AC} + \left((1-\gamma)\frac{\beta-2}{3}+\beta-\gamma\right) n_A n_{ABC} + \\
& + \beta\left(1+\frac{2}{3}(1-\gamma)\right) n_{ABC} n_{AC} + (1-\gamma)\beta\left(1+\frac{2}{3}\right) n_{ABC} n_{AB} + (1-\gamma)\frac{2}{3}\beta n_{ABC}^2
\end{aligned}
$$

$$
\begin{aligned}
\frac{dn_B}{dt} =\ & -n_A n_B - n_C n_B + \left((1-\gamma)\frac{\beta-1}{2}+\beta-\gamma\right) n_{AB} n_B + \frac{3\beta-1}{2} n_{BC} n_B - n_{AC} n_B + \\
& + (1-\gamma)\beta n_{AB}^2 + \beta n_{BC}^2 + \beta\left((1-\gamma)+1\right) n_{AB} n_{BC} + \left((1-\gamma)\frac{\beta-2}{3}+\beta-\gamma\right) n_B n_{ABC} + \\
& + \beta\left(1+\frac{2}{3}(1-\gamma)\right) n_{ABC} n_{BC} + (1-\gamma)\beta\left(1+\frac{2}{3}\right) n_{ABC} n_{AB} + (1-\gamma)\frac{2}{3}\beta n_{ABC}^2
\end{aligned}
$$

$$
\begin{aligned}
\frac{dn_C}{dt} =\ & -n_A n_C - n_B n_C + \frac{3\beta-1}{2}\left(n_{BC}+n_{AC}\right) n_C - (1-\gamma) n_{AB} n_C + \\
& + \gamma n_{AB}\left(2\beta\left(n_{BC}+n_{AC}+n_{ABC}\right)+(1-\beta)\left(n_{ABC}+n_{BC}+n_{AC}\right)+n_A+n_B+n_C+n_{AB}\right) + \\
& + \beta n_{AC}^2 + \beta n_{BC}^2 + 2\beta n_{AC} n_{BC} + \left(\beta+\gamma+(1-\gamma)\frac{\beta}{3}-(1-\gamma)\frac{2}{3}\right) n_{ABC} n_C + \\
& + \gamma n_{ABC}\left(n_A+n_B+n_{AB}\right) + \left(2\beta\gamma+(1-\gamma)\frac{2}{3}\beta+(1-\beta)\gamma+\beta\right) n_{ABC}\left(n_{AC}+n_{BC}\right) + \\
& + \left(\gamma(1-\beta)+2\gamma\beta+(1-\gamma)\frac{2}{3}\beta\right) n_{ABC}^2
\end{aligned}
$$

$$
\begin{aligned}
\frac{dn_{AB}}{dt} =\ & 2n_A n_B + \frac{1}{2} n_{AC} n_B + \frac{1}{2} n_{BC} n_A - (\gamma+1) n_{AB} n_C - \left((1-\gamma)\frac{\beta-1}{2}+\beta+\gamma\right) n_{AB}\left(n_A+n_B\right) + \\
& - 2\left(\beta(1-\gamma)+\gamma\right) n_{AB}^2 - \left(\frac{1+\beta}{2}+\gamma+(1-\gamma)\frac{\beta}{2}\right) n_{AB}\left(n_{AC}+n_{BC}\right) + \\
& + \frac{1-\gamma}{3} n_{ABC}\left(n_A+n_B\right) - \left((1-\gamma)\frac{2}{3}\beta+2\gamma+\frac{1-\gamma}{3}+(1-\gamma)\beta\right) n_{ABC} n_{AB}
\end{aligned}
$$

$$
\begin{aligned}
\frac{dn_{AC}}{dt} =\ & 2n_A n_C - 2\beta n_{AC}^2 - \frac{3\beta-1}{2} n_{AC}\left(n_A+n_C\right) + \frac{1-\gamma}{2} n_{AB} n_C + \frac{1}{2} n_{BC} n_A + \gamma n_{AB} n_A - n_B n_{AC} + \\
& - \left(\beta\left(\gamma+\frac{1-\gamma}{2}+\frac{1}{2}\right)+\frac{1-\gamma}{2}\right) n_{AB} n_{AC} - \frac{2\beta+1}{2} n_{AC} n_{BC} + \left(\gamma+\frac{1-\gamma}{3}\right) n_{ABC} n_A + \\
& + \frac{1-\gamma}{3} n_{ABC} n_C - \left(\beta\left(\gamma+1+\frac{2}{3}(1-\gamma)\right)+\frac{1-\gamma}{3}\right) n_{AC} n_{ABC}
\end{aligned}
$$

$$
\begin{aligned}
\frac{dn_{BC}}{dt} =\ & 2n_B n_C - 2\beta n_{BC}^2 \frac{3\beta-1}{2} n_{BC}\left(n_B+n_C\right) + \frac{1-\gamma}{2} n_{AB} n_C + \frac{1}{2} n_{AC} n_B + \gamma n_{AB} n_B - n_A n_{BC} + \\
& - \left(\beta\left(\gamma+\frac{1-\gamma}{2}+\frac{1}{2}\right)+\frac{1-\gamma}{2}\right) n_{AB} n_{BC} - \frac{2\beta+1}{2} n_{BC} n_{AC} + \left(\gamma+\frac{1-\gamma}{3}\right) n_{ABC} n_B + \\
& + \frac{1-\gamma}{3} n_{ABC} n_C - \left(\beta\left(\gamma+1+\frac{2}{3}(1-\gamma)\right)+\frac{1-\gamma}{3}\right) n_{BC} n_{ABC}
\end{aligned}
$$

$$
\begin{aligned}
\frac{dn_{ABC}}{dt} =\ & n_A n_{BC} + n_B n_{AC} + n_C n_{AB} + \frac{1}{2}\left(2 n_{BC} n_{AC}+(1+(1-\gamma)) n_{AB}\left(n_{AC}+n_{BC}\right)\right) + \\
& + \frac{1-\gamma}{3} n_{ABC}\left(n_{AB}+n_{AC}+n_{BC}\right) + \gamma n_{AB}^2 - \beta\left(n_A+n_B+n_C\right) n_{ABC} + \\
& - \beta\left((1-\gamma) n_{AB}+n_{AC}+n_{BC}\right) n_{ABC} - \frac{1-\gamma}{3}\beta n_{ABC}\left(n_A+n_B+n_C\right) + \\
& - \frac{2(1-\gamma)}{3}\beta n_{ABC}\left(n_{AB}+n_{AC}+n_{BC}\right) - \gamma\beta n_{AB} n_{ABC} - \gamma n_{ABC}\left(n_A+n_B+n_C\right) + \\
& - \gamma n_{ABC}\left(n_{AC}+n_{BC}\right) - 2(1-\gamma)\beta n_{ABC}^2 - 2\gamma\beta n_{ABC}^2 - \gamma(1-\beta) n_{ABC}^2
\end{aligned} \tag{1}
$$

# References

1. Castellano, C., Fortunato, S., Loreto, V.: Statistical physics of social dynamics. Rev. Mod. Phys. 81, 591–646 (2009)
2. Loreto, V., Baronchelli, A., Mukherjee, A., Puglisi, A., Tria, F.: Statistical physics of language dynamics. J. Stat. Mech. P04006 (2011)
3. Steels, L.: A self-organizing spatial vocabulary. Artificial Life 2, 319–332 (1995)
4. Satterfield, T.: Toward a Sociogenetic Solution: Examining Language Formation Processes Through SWARM Modeling. Social Science Computer Review 19, 281–295 (2001)
5. Nakamura, M., Hashimoto, T., Tojo, S.: Prediction of creole emergence in spatial language dynamics. In: Dediu, A.H., Ionescu, A.M., Martín-Vide, C. (eds.) LATA 2009. LNCS, vol. 5457, pp. 614–625. Springer, Heidelberg (2009)
6. Strimling, P., Parkvall, M., Jansson, F.: Modelling the evolution of creoles. In: The Evolution of Language: Proceedings of the 9th International Conference (EVOLANG9), pp. 464–465. World Scientific Publishing Company (2012)
7. Baronchelli, A., Dall'Asta, L., Barrat, A., Loreto, V.: Nonequilibrium phase transition in negotiation dynamics. Phys. Rev. E 76, 051102 (2007)
8. Mufwene, S.S.: The ecology of language evolution. Cambridge University Press, Cambridge (2001)
9. Steels, L.: Modeling the cultural evolution of language. Physics of Life Reviews 8, 339–356 (2011)
10. Mufwene, S.S.: Population movements and contacts in language evolution. Journal of Language Contact-THEMA 1 (2007)

# Trust-Based Scenarios – Predicting Future Agent Behavior in Open Self-organizing Systems

Gerrit Anders, Florian Siefert, Jan-Philipp Steghöfer, and Wolfgang Reif

Institute for Software and Systems Engineering, Augsburg University, Germany
{anders,siefert,steghoefer,reif}@informatik.uni-augsburg.de

**Abstract.** Agents in open self-organizing systems have to cope with a variety of uncertainties. In order to increase their utility and to ensure stable operation of the overall system, they have to capture and adapt to these uncertainties at runtime. This can be achieved by formulating an expectancy of the behavior of others and the environment. Trust has been proposed as a concept for this purpose.

In this paper, we present trust-based scenarios as an enhancement of current trust models. Trust-based scenarios represent stochastic models that allow agents to take different possible developments of the environment's or other agents' behavior into account. We demonstrate that trust-based scenarios significantly improve the agents' capability to predict future behavior with a distributed power management application.

**Keywords:** Scenarios, Trust, Robustness, Resilience, Uncertainty, Self-Organizing Systems, Adaptive Systems, Open Multi-Agent Systems.

## 1 Uncertainties in Open Self-Organizing Systems

In open self-organizing systems, agents interact with and are embedded in a heterogeneous and dynamic environment in which potential interaction partners might not cooperate [14]. These characteristics introduce uncertainty into the decision making process of each individual agent. But even in a cooperative environment, uncertainties are relevant since agents usually have imperfect and limited knowledge about their environment or the behavior of other agents. Agents that intentionally cheat to increase their own utility as well as agents that unintentionally do not behave as expected can not only compromise the agents' efficiency but also their ability to effectively pursue and reach their objectives. The problem is exacerbated if the regarded systems are mission-critical. In such cases, the resilience and dependability of the systems hinge on the ability to deal with uncertainties introduced by other agents and the environment.

Decentralized power management systems are an example of mission-critical systems. Because their failure can have massive consequences for people, industries, and public services, it is of utmost importance that they are stable and available at all times. However, they have to deal with a variety of uncertainties introduced by the systems' participants and their environment. A dependable and resilient control scheme is thus necessary that, at the same time, must be

W. Elmenreich, F. Dressler, and V. Loreto (Eds.): IWSOS 2013, LNCS 8221, pp. 90–102, 2014.

able to adapt to changing conditions quickly and autonomously. In [1], we introduced the concept of Autonomous Virtual Power Plants (AVPPs) that embodies such a control scheme in the context of a self-organizing system structure. Each AVPP controls a group of power plants. The structure changes in response to new information and changing conditions to enable each AVPP to balance its power demand and production. For this purpose, each AVPP calculates schedules that stipulate the output of controllable power plants for future points in time. The scheduled output of controllable power plants has to satisfy the "residual load" which arises as the difference between the power demand and the production of the non-controllable power plants like weather-dependent generators. To approximate the residual load, AVPPs have to rely on predictions about the future demand as well as the future output of non-controllable power plants. However, these predictions are subject to uncertainties: consumers and power plants can behave arbitrarily since they pursue individual goals and are exposed to their environment, resulting in (un)intentionally inaccurate predictions; the behavior is variable since it depends on external conditions (e.g., if snow covers the solar panels); and the behavior is time-dependent since prior good predictions are usually an indicator for future good ones. If these uncertainties are known and thus a model of the consumers' and power plants' underlying stochastic processes is available, an AVPP can use demand and power predictions to predict the corresponding predictor's expected behavior. Incorporating uncertainties into the schedule creation then allows AVPPs to prevent or mitigate imbalances between energy production and demand that are caused by inaccurate predictions.

As a way to deal with uncertainties, trust models have been proposed (see Sect. 2). They allow agents to measure and quantify other agents' behavior by means of a trust value. The value captures the experiences with that particular agent and allows to derive whether or not the interactions where beneficial in the past. If a trust value is regarded as a measure of uncertainty, it can in turn be used to make evidence-based assumptions about an agent's future behavior. However, in general, trust values do not capture the stochastic process underlying the observed behavior. This information is lost when the trust metric derives the value. Instead, an average over the last experiences is formed that, at best, allows to predict with which probability a future interaction will again be beneficial.

In the domain of operations research, *scenarios* are a well-known concept to mirror a system's underlying stochastic process [5]. It is expressed by different possible developments of the system. As each scenario has a certain probability of occurrence, an agent can, e.g., choose the most likely scenario and optimize for the future the scenario predicts. However, current approaches usually use predetermined scenarios and probabilities [15]. In an open self-organizing system, it is not possible to determine possible scenarios beforehand. As self-interested agents come and go and interaction patterns change, no assumptions can be made about the individual agent's behavior. It is thus essential that scenarios and their probabilities are determined at runtime with up-to-date data.

These characteristics imply several challenges that an approach to measure uncertainties and predict future behavior has to deal with:

1. **Arbitrary behavior:** No assumptions about agent behavior can be made. The stochastic model must depict behavior that can be described by continuous or discrete random variables as well as different probability distributions.
2. **Subjective behavior:** Because agents can behave differently towards different interaction partners, each agent has to build its individual stochastic model of others.
3. **Variable behavior:** An agent's behavior can vary and completely change over time as it can adapt to changes in its environment and, e.g., adjusts its objectives accordingly. The stochastic model must be flexible enough to reflect these changes, e.g., by a mechanism that allows for forgiveness.
4. **Time-dependent behavior:** An agent's current behavior might depend on its behavior in the past. A stochastic model should thus be able to mirror time-dependent behavior.

In this paper, we propose to meet these challenges by dynamically calculating *trust-based scenarios* and their probabilities at runtime. They provide a stochastic model of sources of uncertainty, allowing agents to predict and deal with them by making informed and robust decisions. With regard to AVPPs, i.e., our running example, this increases the system's resilience against mispredictions. Sect. 2 gives an overview of existing trust models, while Sect. 3 introduces trust-based scenarios. Our running example is then used to evaluate our approach in Sect. 4. In Sect. 5, we conclude the paper and give an outlook on future work.

## 2    Advantages and Limitations of Current Trust Models

*Trust* is a multi-faceted concept that allows agents to appraise their interaction partners' behavior. Among others, trust includes the facets reliability and credibility [14]. An agent's reliability specifies its availability, while an agent's credibility indicates its willingness to participate in an interaction in a desirable manner and corresponds to the original notion of trust in MAS [12]. We focus on credibility in this paper. In principle, an agent $a$'s trust in an interaction partner $b$ results from *experiences* with $b$. Each gathered experience stems from a *contract* in which the desired result of $a$'s interaction with $b$ was stipulated. The experience is created after the interaction is completed. It contains the stipulated and the actual result of the interaction. Trust is highly subjective as an agent can show different behavior towards its interaction partners. Because of an agent's capabilities and objectives, trust further depends on the *context*, e.g., the environmental circumstances, in which an interaction takes place. For the sake of simplicity, we deliberately abstain from trust context considerations here.

If it is valid to assume that an agent's prior behavior is indicative for its future behavior, trust can be used to predict this future behavior and is thus a measure to cope with uncertainties as shown in Sect. 2.1 and Sect. 2.2.

## 2.1   Predicting Agent Behavior by Means of Trust Values

Because an interaction can last multiple time steps, we define a contract $C_t^i = (c_{t+0}^i, ..., c_{t+n}^i)$ as an (n+1)-tuple that comprises multiple stipulated results $c_{t+j}^i$, where $j, n, t \in \mathbb{N}_0^+$, $i \in \mathbb{N}^+$ is a unique identifier, and $t + j$ identifies the time step in which the interaction partner should behave as stated in $c_{t+j}^i$. $t$ and $t + n$ thus specify the time frame in which $C_t^i$ is valid. With respect to power, i.e., residual load, predictions in AVPPs (see Sect. 1), $n + 1$ is the length of the prediction. For instance, if the power prediction covers a time frame of 8 hours in which the residual load is predicted in 15 minute intervals, we have $n + 1 = 32$. In the following, let $[X]_j$ denote the $j$-th element of a tuple $X$. An atomic experience $e_{t+j}^i = (c_{t+j}^i, r_{t+j}^i)$ is a 2-tuple, consisting of the stipulated result $c_{t+j}^i$ and the actual result $r_{t+j}^i$. An atomic experience $e_{t+j}^i = (7 \text{ MW}, 8 \text{ MW})$ with an AVPP's residual load states that a residual load of 7 MW was stipulated for time step $t + j$, but 8 MW were measured. Consequently, an experience $E_t^i = (e_{t+0}^i, ..., e_{t+n}^i)$ is an (n+1)-tuple of atomic experiences $e_{t+j}^i = [E_t^i]_j$, and $t + j$ is the time step in which $[E_t^i]_j$ was gained. Contracts $C_t^i$ and experiences $E_t^i$ comprise $n + 1$ so-called *time slots*, e.g., $[E_t^i]_j$ was gained in the $j$-th time slot.

If an agent $a$ evaluates the trustworthiness of an agent $b$, it uses a *trust metric* $\mathcal{M} : \mathcal{E} \times ... \times \mathcal{E} \to \mathcal{T}$ to evaluate a number of experiences with $b$ ($\mathcal{E}$ is the domain of experiences). The metric returns a *trust value* $\tau \in \mathcal{T}$ and relies on a *rating function* $\mathcal{R} : \epsilon \to \mathcal{T}$ that appraises atomic experiences ($\epsilon$ is the domain of atomic experiences). The result of $\mathcal{R}$ is a rating $\pi \in \mathcal{T}$. $\mathcal{T}$ usually is an interval $[0, 1]$ or $[-1, 1]$. Regarding $\mathcal{T} = [0, 1]$, a trust value $\tau = 0$ or $\tau = 1$ states that agent $b$ either never or always behaves beneficially [10]. However, $b$ behaves predictably in both cases. If the trust value is around the interval's midpoint, $b$'s behavior is highly unpredictable and thus induces a high level of uncertainty.

Because the residual load can be over- or underestimated, we use $\mathcal{T} = [-1, 1]$ so that positive and negative deviations from predictions can be captured. A rating $\pi = 0$ states that the residual load is predicted exactly, whereas $\pi = -1$ or $\pi = 1$ state that the residual load is greatly under- or overestimated (i.e., the actual residual load is far higher or lower than predicted).

A trust value has to be semantically sound to allow valid predictions of an agent's future behavior. This property depends on the metric $\mathcal{M}$. $\mathcal{M}$ can, e.g., calculate the mean deviation between the stipulated $c_{t+j}^i \in \mathbb{R}$ and actual result $r_{t+j}^i \in \mathbb{R}$ of atomic experiences $[E_{t_h}^{i_h}]_j$ contained in a list of $m$ experiences $E_{t_1}^{i_1}, ..., E_{t_m}^{i_m}$ ($k \in \mathbb{R}$ equals the maximum possible or, if not available, observed deviation from a contract and thus normalizes the result to a value in $[-1, 1]$):

$$\mathcal{M}(E_{t_1}^{i_1}, ..., E_{t_m}^{i_m}) = \frac{\sum_{h=1}^m \sum_{j=0}^n \mathcal{R}([E_{t_h}^{i_h}]_j)}{m \cdot (n + 1)}; \quad \mathcal{R}([E_t^i]_j) = \frac{c_{t+j}^i - r_{t+j}^i}{k} \quad (1)$$

Based on $\mathcal{M}$, a trust value $\tau$, and a contract $C_t^i$, an agent can predict the *expected behaviors* $B_t^i = (b_{t+0}^i, ..., b_{t+n}^i)$ of its interaction partner during $C_t^i$'s validity. With respect to Eq. 1, the agent's *expected behavior* $[B_t^i]_j$ in time step $t+j$ is defined as the difference between $[C_t^i]_j$ and the expected deviation $\tau \cdot k$:

$$[B_t^i]_j = [C_t^i]_j - \tau \cdot k \tag{2}$$

For example, if $k = 10$ MW, $\tau = 0.1$, and the power prediction's stipulated results are $C_t^i = (5 \text{ MW}, 6 \text{ MW})$, the expected residual load can be predicted as $B_t^i = (4 \text{ MW}, 5 \text{ MW})$. If the AVPP schedules its subordinate controllable power plants on the basis of $B_t^i$ instead of $C_t^i$, it is expected that the deviation between the power plants' output and the actual residual load can be decreased. However, the prediction of the residual load's future behavior with Eq. 2 can be imprecise because we disregard that an agent's behavior can be arbitrary and that it might be time-dependent (see Sect. 1). Since agents can behave arbitrarily, one and the same trust value can stem from very different experiences, e.g., $\tau = 0.1$ could be based on experiences in which the residual load was always 1 MW lower than stipulated or a situation in which 25% of the predictions were overestimated by 2 MW and 75% of the predictions were underestimated by $-2$ MW. With regard to time-dependent behavior, the prediction of the residual load for a time step $t$ could, e.g., tend to be rather precise if the prediction for the previous time step $t - 1$ is accurate. A similar dependence could exist for inaccurate predictions.

As a trust value is usually just a mean value, it is not applicable to predict an agent's behavior in this case. Regarding an agent's underlying stochastic process, which can be represented by a probability distribution, the mean value simply does not capture enough information to approximate and describe it sufficiently.

## 2.2    Other Trust Models to Predict Agent Behavior

In the body of literature, various trust models have been presented that can be used to predict the environment's or other agents' behavior in specific situations.

In [8], confidence is proposed as a concept that indicates the degree of certainty that a trust value mirrors an agent's actual observable behavior. Among others, it depends on the standard deviation of experiences used to assess the trust value. The combined information given by the trust value (i.e., the mean behavior) and the standard deviation is still insufficient to approximate an arbitrary stochastic process underlying an agent's behavior though. The concrete probability distribution an agent's behavior follows must still be assumed.

In [6], an agent's trust value in the next time step is predicted on the basis of a Markov model. States mirror an agent's trustworthiness. The probability associated with a state change is determined at runtime, dependent on how often this transition was observed. While this model includes a basic mechanism to reflect time-dependent behavior (the trust value in the next time step depends on the current trust value), it is not applicable in situations that need predictions for more than a single time step or in situations in which too few data is available (e.g., there might be no transitions for the current trust value).

To be able to consider (in)regularities when predicting future trust values (such a behavior is shown by strategic agents), [7] presents several metrics that can determine the trust value for the next time step dependent on such properties. However, the applicability of the metrics depends on the characteristic of the agent's behavior and it is not possible to predict different possible developments of the trust value. Apart from this, these metrics basically exhibit similar drawbacks as the Markov chain method presented in [6].

Another trust model that is based on a Markov chain is introduced in [3]. In a mobile ad-hoc network, events, i.e., new experiences, trigger state changes in the Markov chain, i.e., changes in the trust value. While the Markov chain is used to analyze the trust model, it remains unclear how agents use information like transition probabilities to make decisions or predict future behavior.

Finally, [9] presents a fuzzy regression-based approach to predict uncertainties in the field of service-oriented applications. Predictions are made on the basis of a trust value, a trust trend, and the performance consistency that reflects the trust value's stability. While this model can predict behavioral changes and trends in an e-service environment, the stochastic model an agent can derive from it can not be applied in the domain of open self-organizing systems as the challenges mentioned in Sect. 1 are not completely met.

## 3   Trust-Based Scenarios

Scenarios are a proven concept to approximate a stochastic process by taking several possible developments of a system into account (see, e.g., [11]). These developments are often represented by a *scenario tree* which can be annotated with probabilities that the system develops in a specific direction. Scenario trees then serve as input for solving optimization problems under uncertainty such as stochastic programming [13]. These techniques are applied, e.g., in the domain of power management systems (see, e.g., [4,2,15]). In literature, scenarios are generated, e.g., by solving an optimization problem on the basis of a system model, gathered historical data, and expert knowledge [5]. The scenario tree's structure, the number of scenarios, or probability distributions are often predefined (see, e.g., [4,2]). Due to the computational complexity, scenarios are often determined off-line [15]. In [15], a Markov process is introduced as a simple mechanism to generate scenarios as input to solve a power plant scheduling problem. However, the authors act on the assumption that the scenarios and transition probabilities are predetermined at design time and not adapted at runtime. Further, simple scenario generation mechanisms, like the one presented in [15], often lack the ability to mirror time-dependent behavior. Scenario generation is thus often based on assumptions that do not meet the challenges outlined in Sect. 1.

In the following, we present an approach to generate trust-based scenarios (TBSs) at runtime which can in turn be used to predict the environment's or an agent's behavior on the basis of experiences gained in prior interactions.

### 3.1   Generating Trust-Based Scenarios

Our approach to generate TBSs is similar to the creation of a histogram or a Markov chain with variable transition probabilities. We assume that an agent's behavior can be assessed by a rating function like the one given in Eq. 1 and that an agent's prior behavior is indicative of its future behavior. Basically, we categorize atomic experiences according to their rating into classes of equivalent behavior and thus approximate an agent's underlying stochastic process. The TBSs originate from experiences that have been identically classified and serve as input to generate a trust-based scenario tree (TBST). If we have contracts and experiences of length $n + 1$, an agent can predict the behavior of an interaction partner or its environment in up to $n + 1$ future time steps. The probability that an agent changes its behavior from one class to another depends on the number of experiences in which this transition was actually observed. In the following, we explain in more detail how TBSs are generated.

**Classifying Experiences:** As mentioned above, we categorize rated atomic experiences into classes. For each regarded time slot $j \in \{0, ..., n\}$, these classes are consecutive and non-overlapping intervals $\Delta\pi_j \in \Delta\mathcal{T}_j$ that only contain elements in $\mathcal{T}$ so that each possible rating $\pi$ can be assigned to an interval, i.e., $\forall \pi \in \mathcal{T} : \exists \Delta\pi_j \in \Delta\mathcal{T}_j : \pi \in \Delta\pi_j$. Suitable sizes of these intervals depend on the application. To mirror time-dependent behavior, there is a set $\Delta\mathcal{T}_j$ for each time slot $j$ so that an atomic experience $[E_t^i]_j$ is classified into an interval $\Delta\pi_j \in \Delta\mathcal{T}_j$.

By assigning each rated atomic experience $\mathcal{R}([E_t^i]_j)$ contained in an experience $E_t^i \in \mathcal{E}$ to such an interval $\Delta\pi_j \in \Delta\mathcal{T}_j$, $E_t^i$ is converted into a *classified experience* $\Xi^i \in \mathcal{E}^*$ ($\mathcal{E}^*$ denotes the set of classified experiences). A classified experience $\Xi^i = (\Delta\pi_0, ..., \Delta\pi_n)$ is thus an $(n+1)$-tuple, i.e., a sequence, of intervals $\Delta\pi_j \in \Delta\mathcal{T}_j$ that contain the corresponding rated atomic experience $\mathcal{R}([E_t^i]_j)$ of $E_t^i$ (see Eq. 3). In Fig. 1, steps 1 and 2 illustrate how experiences are classified.

$$\Xi^i :\Leftrightarrow \forall j \in \{0, ..., n\} : \mathcal{R}([E_t^i]_j) \in [\Xi^i]_j \wedge [\Xi^i]_j \in \Delta\mathcal{T}_j \qquad (3)$$

**Determining Trust-Based Scenarios:** A *trust-based scenario* (TBS) $\Xi = (\Delta\pi_0, ..., \Delta\pi_n)$ arises from classified experiences $\Xi^i = (\Delta\pi_0, ..., \Delta\pi_n)$. More precisely, each group of experiences that were classified into the same sequence of intervals forms a TBS (the set $\mathcal{S} = \Delta\mathcal{T}_0 \times ... \times \Delta\mathcal{T}_n$ contains all possible TBSs). Note that in contrast to $\Xi^i$, $\Xi$ has no identifier $i$. For each $\Xi$, a probability of occurrence $p^\Xi$ can be calculated. It is the ratio of the number of classified experiences $\Xi^i$ that feature the same sequence of intervals as defined by $\Xi$ ($\forall j \in \{0, ..., n\} : [\Xi^i]_j = [\Xi]_j$) to the number of all classified experiences $|\mathcal{E}^*|$:

$$p^\Xi := \frac{1}{|\mathcal{E}^*|} \cdot \left| \left\{ \Xi^i \, \middle| \, \forall \Xi^i \in \mathcal{E}^* : \forall j \in \{0, ..., n\} : [\Xi^i]_j = [\Xi]_j \right\} \right| \qquad (4)$$

Hence, each generated TBS $\Xi$ has a probability $p^\Xi > 0$. These TBSs form the set $\mathcal{S}^+ \subseteq \mathcal{S}$. If $\Xi$ is used to predict an agent's behavior, it thus represents a corridor of expected behavior. In Fig. 1, steps 2 and 3 show how TBSs are created.

**Deriving a Trust-Based Scenario Tree:** The TBSs can now be used to derive an agent's stochastic model, the *trust-based scenario tree* $TBST = (\Theta, \Phi)$, where $\Theta$ is the set of nodes and $\Phi$ the set of edges. The TBST is based on all generated TBSs (i.e., all $\Xi \in \mathcal{S}^+$). A node $\theta^\rho \in \Theta$ is defined by an interval $\theta$ and a sequence of intervals $\rho$. Let $[\Xi]^j = ([\Xi]_0, ..., [\Xi]_{j-1})$ specify the first $j$

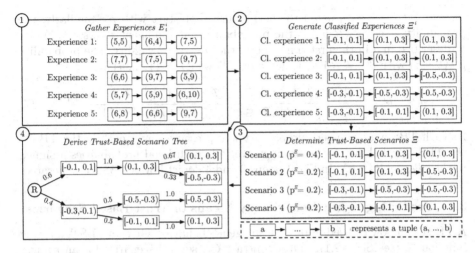

**Fig. 1.** Trust-based scenarios approximate an agent's underlying stochastic process by classifying gathered experiences, determining trust-based scenarios (TBSs), and thereupon deriving a trust-based scenario tree (TBST). The TBST is then used to predict future behavior on the basis of a stipulated contract. In this example, $\mathcal{T} = [-1, 1]$ and $\forall j \in \{0, ..., n\} : \Delta\mathcal{T}_j = \{[-1.0,-0.9),[-0.9,-0.7),...,[-0.3,-0.1),[-0.1,0.1],(0.1,0.3],$ ..., $(0.7,0.9],(0.9,1.0]\}$. Atomic experiences are rated using the rating function given in Eq. 1 with $k = 10$. The length of experiences is $n+1 = 3$. Tuples $(a, b)$ represent atomic experiences $[E_i^i]_j$. In step 4, the values at the TBST's edges indicate the conditional probabilities to change from one behavior to another.

elements $[\Xi]_j$ of the interval sequence defined by a TBS $\Xi$, i.e., a prefix of length $j$ (if $j < 1$, $[\Xi]^j$ is the empty sequence "$\emptyset$"). Then, for each $j \in \{0, ..., n\}$ and each TBS $\Xi \in \mathcal{S}^+$, there is a node $\theta^\rho \in \Theta$ with $\theta = [\Xi]_j$ and $\rho = [\Xi]^j$. A node $\theta^\rho \in \Theta$ thus states that an agent showed a behavior classified as $\theta$ after having shown behaviors classified as $\rho$ in the given order. Additionally, $\Theta$ contains the tree's root $R$. If there are two nodes $\theta_1^{\rho_1}$ and $\theta_2^{\rho_2}$ with $\rho_1 = [\Xi]^{j-1}$ and $\rho_2 = [\Xi]^j$, there is an edge $(\theta_1^{\rho_1}, \theta_2^{\rho_2}) \in \Phi$. Note that $\rho_2$ is equal to the concatenation $\rho_1 \oplus \theta_1$. Further, for each TBS $\Xi$, $\Phi$ contains an edge $(R, \theta^\emptyset)$ from $R$ to the interval $\theta = [\Xi]_0$ at time slot 0. More precisely, nodes and edges are defined as follows:

$$\Theta := R \cup \{\theta^\rho \,|\, \exists \Xi \in \mathcal{S}^+ : \exists j \in \{0, ..., n\} : \theta = [\Xi]_j \wedge \rho = [\Xi]^j\} \tag{5}$$

$$\Phi := \{(R, \theta^\rho) \,|\, \exists \Xi \in \mathcal{S}^+ : \theta = [\Xi]_0 \wedge \rho = \emptyset\} \cup \{(\theta_1^{\rho_1}, \theta_2^{\rho_2}) \,|\, \exists \theta_1^{\rho_1}, \theta_2^{\rho_2} \in \Theta : \tag{6}$$
$$\exists \Xi \in \mathcal{S}^+ : \exists j \in \{1, ..., n\} : \rho_1 = [\Xi]^{j-1} \wedge \rho_2 = [\Xi]^j\}$$

As all leafs are at depth $n+1$, each path from the TBST's root to a leaf represents a TBS, and each TBS with non-zero probability is represented by such a path.

In a TBST, each edge $(\theta_1^{\rho_1}, \theta_2^{\rho_2})$ has a probability $p(\theta_1^{\rho_1}, \theta_2^{\rho_2})$ that indicates the conditional probability that the agent will show a behavior classified as $\theta_2$ if it showed behaviors specified by $\rho_2$ in the previous time steps. This probability is the ratio of the number of classified experiences that start with the sequence

$\rho_1 \oplus \theta_1 \oplus \theta_2$ ($\rho_2 = \rho_1 \oplus \theta_1$) to the number that start with $\rho_2$. Similarly, the probability of edges $(R, \theta^\emptyset)$ is the probability that the agent shows a behavior as described in $\theta$ in the first time step of the contract's validity. The probability $p(\theta_1^{\rho_1}, \theta_2^{\rho_2})$ with $\theta_1^{\rho_1} \neq R$ is calculated as follows ($\mathcal{J} = \{1, ..., n\}$):

$$p(\theta_1^{\rho_1}, \theta_2^{\rho_2}) := \frac{|\{\Xi^i \,|\, \exists \Xi^i \in \mathcal{E}^* : \exists j \in \mathcal{J} : \rho_2 = [\Xi^i]^j \wedge \theta_2 = [\Xi^i]_j \}|}{|\{\Xi^i \,|\, \exists \Xi^i \in \mathcal{E}^* : \exists j \in \mathcal{J} : \rho_2 = [\Xi^i]^j \}|} \qquad (7)$$

As each TBS with non-zero probability is represented by a path in the TBST, the TBS's probability is the product over all conditional probabilities assigned to the edges of its path. In Fig. 1, steps 3 and 4 depict how a TBST is generated.

## 3.2   Predicting Agent Behavior by Means of Trust-Based Scenarios

A TBST can now be used to predict an agent's behavior. Given a TBST, a rating function $\mathcal{R}$ (see Sect. 2.1), and a contract $C_t^i$, several scenarios for an agent's expected behavior can be determined. For this purpose, each interval $\theta = [\overline{\theta}, \underline{\theta}]$ that is represented by a node $\theta^\rho$ in the TBST with lower and upper endpoints $\overline{\theta}$ and $\underline{\theta}$ is transformed into a *corridor of expected behavior* $[\overline{b_j}, \underline{b_j}]$, similar to Eq. 2. If $\mathcal{R}$ and $k$ are defined as in Eq. 1, this is done as follows:

$$[\overline{b_j}, \underline{b_j}] = [([C_t^i]_j - \overline{\theta} \cdot k), ([C_t^i]_j - \underline{\theta} \cdot k)] \qquad (8)$$

Consequently, the TBST reflects different possible future developments of an agent's or the environment's behavior. Obviously, the smaller the size of the intervals $\Delta \pi_j$ used to classify the atomic experiences, the narrower these corridors and the higher the accuracy of predicted behavior. However, this accuracy heavily depends on the number of gathered experiences the TBST is based on.

To make decisions, an agent can take the whole TBST, a subtree, or a single TBS into account. In case a single TBS is preferred, the agent could choose the TBS represented by a path that is obtained by, starting at the root, choosing the edge with the highest conditional probability. This TBS is not necessarily the most likely TBS, but, under the assumption that it is more important to predict an agent's behavior in the near future than in the distant future, this procedure yields very precise predictions (see Sect. 4).

As agents might show time-dependent behavior, predictions can be improved by selecting relevant TBSs on the basis of an agent's behavior in recent time steps. The TBST is then derived from these TBSs. For example, if the accuracy of the prediction of the residual load depends on the accuracy of predictions of the last $x$ time steps, then it would be beneficial to take the prediction quality of the last $x$ time steps into account when predicting the expected residual load's behavior by means of TBSs. This is achieved as follows: Whenever a contract $C_t^i$ is concluded, the last $x$ atomic experiences gained before the current time step are recorded. Regarding the experience $E_t^i$ gained for $C_t^i$ as a list, these recorded atomic experiences are then added to the front of $E_t^i$. When a TBST is generated to predict an agent's expected behavior during the validity of a contract, its behavior in the last $x$ time steps is then used to select a subset of relevant

TBSs from the set of TBSs with non-zero probability. Relevant TBSs are those whose intervals' midpoint in the first $x$ time slots is closest to the rated atomic experiences in the last $x$ time steps. In the course of deriving the TBST from the selected TBSs, these additional atomic experiences are ignored. We show in Sect. 4 that this selection process yields very good results.

## 4   Evaluation

For evaluation, we regarded a setting of a single AVPP and an agent, called *predictor*, that represented all consumers as well as the AVPP's subordinate stochastic power plants. Power plant models and power demand are based on real-world data. In each time step, the predictor predicted the residual load the AVPP had to satisfy in the next $n + 1 = 32$ time steps (a single time step represented 15 minutes). For each time slot, the residual load prediction was created by adding a generated prediction error to the actual residual load. Therefore, in each time step, the predictor generated 32 prediction errors at random, using a gaussian distribution. To reflect time-dependent behavior, prediction errors depended on previous prediction errors. The AVPP's objective was to minimize the deviation between the actual residual load and the residual load it expected. To predict the expected residual load, the AVPP used TBSs and a trust value for comparison. The TBSs and the trust value were determined on the basis of the last $m = 50$ experiences (see Eq. 1) because we identified in preliminary tests that $m = 50$ allows the trust value to achieve good results. The AVPP rated experiences and determined the trust value by means of Eq. 1 with $\mathcal{T} = [-1, 1]$ and $k = 9125$ kW (the AVPP's maximum output/residual load). The AVPP preselected relevant TBSs on the basis of the prediction error of the last two time steps and selected a single TBS to predict the expected behavior as explained in Sect. 3.2. For each parametrization, we performed 100 simulation runs over 1000 time steps. Fig. 2(a) depicts the predictor's mean behavior over time. Fig. 2(b) shows the mean prediction error for each of the 32 time slots of a prediction.

The predictions of the expected residual load are significantly improved when using TBSs instead of a trust value (see Tab. 1 and Fig. 2(a)). Compared to the situation in which the AVPP relied on the residual load prediction, the mean

**Table 1.** Deviations between the actual and the expected residual load: the expected residual load is 1) equivalent to the predictor's residual load prediction ("unmodified prediction"), 2) based on a trust value (see Sect. 2.1), 3) based on TBSs (see Sect. 3).

| Predicted Behavior | Number of Classes | Size of Classes | $\mu_\Delta$ | $\sigma_\Delta$ | $min_\Delta$ | $max_\Delta$ |
|---|---|---|---|---|---|---|
| 1) Unmodified Prediction | - | - | 629.1 | 32.8 | 573.6 | 707.3 |
| 2) Trust Value | - | - | 188.9 | 34.9 | 121.2 | 273.2 |
| 3) Trust-Based Scenarios | $\|\Delta \mathcal{T}_j\| = 365$ | $\approx 50$ kW | 100.4 | 14.6 | 69.0 | 160.6 |
| | $\|\Delta \mathcal{T}_j\| = 183$ | $\approx 100$ kW | 101.7 | 14.0 | 73.0 | 152.5 |
| | $\|\Delta \mathcal{T}_j\| = 91$ | $\approx 201$ kW | 109.0 | 14.7 | 76.5 | 165.6 |
| | $\|\Delta \mathcal{T}_j\| = 45$ | $\approx 405$ kW | 130.2 | 17.5 | 91.6 | 185.0 |

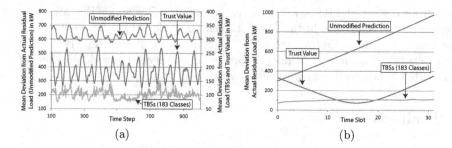

**Fig. 2.** Mean deviation of the expected residual load (unmodified residual load prediction, trust value, TBSs) from the actual residual load. Fig. 2(a) depicts this deviation over time. Fig. 2(b) shows the accuracy of the expected behavior for the 32 time slots.

deviation $\mu_\Delta$ between the expected and the actual residual load can be reduced by approximately 70% when using trust values compared to approximately 84% when using TBSs with $\forall j \in \{0, ..., n\} : |\Delta \mathcal{T}_j| = 183$. TBSs thus reduce the trust value's $\mu_\Delta$ by 46%. In the power grid, it is of utmost importance to reduce the maximum deviation $max_\Delta$ between the expected and actual residual load. Trust values reduce $max_\Delta$ by 61%. TBSs obtain 78% and thus $max_\Delta$ is 44% lower than the trust value's $max_\Delta$. While we expected that TBSs benefit from a high number of classes $|\Delta \mathcal{T}_j|$, one can see in Tab. 1 that 183 classes were sufficient to perceive the underlying stochastic process. Fig. 2(b) shows that TBSs could estimate the behavior in the next 32 time steps much better than a trust value.

Summarizing, compared to trust values, TBSs significantly increased the AVPP's ability to predict the residual load's behavior. Moreover, the risk the AVPP is exposed to, i.e., the maximum deviations, decreases considerably and the variation in prediction quality, i.e., the standard deviation $\sigma_\Delta$ (see Tab. 1), declines. The latter increases the confidence in the predicted expected behavior.

## 5    Conclusion and Future Work

In this paper, we present trust-based scenarios (TBSs) as an instrument to precisely measure, predict, and deal with uncertainties introduced by agents and their environment. Agents generate TBSs by categorizing and grouping experiences gained in the past. Each TBS gives information about an agent's or the environment's possible future behavior and the probability of occurrence. As agents can take several TBSs into account, they are well-equipped to make informed and robust decisions in and to adapt to an uncertain and possibly malevolent environment. In contrast to the majority of current trust models, TBSs make less assumptions about the agent's behavior. TBSs further mirror time-dependent behavior – a prerequisite for coping with strategic agents.

We illustrated our investigations on the basis of a self-organizing power management system that consists of AVPPs. This mission-critical system relies on predictions to ensure stable operation and availability. The evaluation showed that AVPPs benefit from TBSs since TBSs approximate the stochastic process underlying other agents' or the environment's behavior. Compared to the results achieved using trust values, TBSs reduce uncertainties by 46% on average.

In future work, we will devise an algorithm AVPPs use to calculate power plant schedules on the basis of TBSs, apply the concept of TBSs to the trust facet reliability, and investigate in which way TBSs can be used to proactively trigger reorganization in a self-organizing system in order to improve its resilience.

**Acknowledgments.** This research is partly sponsored by the German Research Foundation (DFG) in the project "OC-Trust" (FOR 1085). The authors would like to thank F. Nafz, H. Seebach, and O. Kosak for the valuable discussions.

# References

1. Anders, G., Siefert, F., Steghöfer, J.-P., Seebach, H., Nafz, F., Reif, W.: Structuring and Controlling Distributed Power Sources by Autonomous Virtual Power Plants. In: Proc. of the Power & Energy Student Summit 2010, pp. 40–42 (October 2010)
2. Bouffard, F., Galiana, F.: Stochastic security for operations planning with significant wind power generation. In: Power and Energy Society General Meeting-Conversion and Delivery of Electrical Energy in the 21st Century, pp. 1–11. IEEE (2008)
3. Chang, B., Kuo, S., Liang, Y., Wang, D.: Markov chain-based trust model for analyzing trust value in distributed multicasting mobile ad hoc networks. In: Asia-Pacific Services Computing Conference, pp. 156–161. IEEE (2008)
4. Densing, M.: Hydro-electric power plant dispatch-planning—multi-stage stochastic programming with time-consistent constraints on risk. Dissertation Abstracts International 68(04) (2007)
5. Hochreiter, R., Pflug, G.: Financial scenario generation for stochastic multi-stage decision processes as facility location problems. Annals of Operations Research 152(1), 257–272 (2007)
6. Hussain, F., Chang, E., Dillon, T.: Markov model for modelling and managing dynamic trust. In: 3rd IEEE International Conference on Industrial Informatics, pp. 725–733. IEEE (2005)
7. Hussain, F., Chang, E., Hussain, O.: A robust methodology for prediction of trust and reputation values. In: Proc. of the 2008 ACM Workshop on Secure Web Services, pp. 97–108. ACM (2008)
8. Kiefhaber, R., Anders, G., Siefert, F., Ungerer, T., Reif, W.: Confidence as a Means to Assess the Accuracy of Trust Values. In: Proc. of the 11th IEEE Int. Conf. on Trust, Security and Privacy in Computing and Communications (TrustCom 2012). IEEE (2012)
9. Li, L., Wang, Y., Varadharajan, V.: Fuzzy regression based trust prediction in service-oriented applications. In: Autonomic and Trusted Computing, pp. 221–235 (2009)
10. Mayer, R.C., Davis, J.H., Schoorman, F.D.: An integrative model of organizational trust. The Academy of Management Review 20(3), 709–734 (1995)
11. Pappala, V., Erlich, I.: Power System Optimization under Uncertainties: A PSO Approach. In: Swarm Intelligence Symposium (SIS 2008), pp. 1–8. IEEE (2008)

12. Ramchurn, S., Huynh, D., Jennings, N.: Trust in multi-agent systems. The Knowledge Engineering Review 19(01), 1–25 (2004)
13. Sahinidis, N.V.: Optimization under uncertainty: state-of-the-art and opportunities. Computers & Chemical Engineering 28(6-7), 971–983 (2004)
14. Steghöfer, J.-P., et al.: Trustworthy Organic Computing Systems: Challenges and Perspectives. In: Xie, B., Branke, J., Sadjadi, S.M., Zhang, D., Zhou, X. (eds.) ATC 2010. LNCS, vol. 6407, pp. 62–76. Springer, Heidelberg (2010)
15. Zhang, B., Luh, P., Litvinov, E., Zheng, T., Zhao, F., Zhao, J., Wang, C.: Electricity auctions with intermittent wind generation. In: Power and Energy Society General Meeting, pp. 1–8. IEEE (2011)

# Addressing Phase Transitions in Wireless Networking Optimization

Maria Michalopoulou and Petri Mähönen*

RWTH Aachen University, Institute for Networked Systems
{mmi,pma}@inets.rwth-aachen.de

**Abstract.** The general aim of this paper is to introduce the notion of phase transitions into wireless networking optimization. Although the theory of phase transitions from statistical physics has been employed in optimization theory, phase transitions in the context of optimization of wireless networks have not yet been considered. In wireless networking optimization, given one or more optimization objectives we often need to define mathematically an optimization task, so that a set of requirements is not violated. However, especially recent trends in wireless communications, such as self-organized networks, femto-cellular systems, and cognitive radios, calls for optimization approaches that can be implemented in a distributed and decentralized fashion. Thus we are interested to find utility-based approaches that can be practically employed in a self-organizing network. We argue that phase transitions can be identified and taken appropriately into account in order to eliminate the emergence of undesirable solutions that lie near the point where the phase transition occurs. As an example we present a simple power control problem for a macrocell-femtocell network scenario. We formulate a distributed framework of the problem where we model a phase transition effect by means of a dummy variable in order to exclude solutions lying in the one side of the phase transition.

## 1 Introduction

The concept of optimization – not only in its strict mathematical sense – is fundamental in wireless networking research. Especially the trend towards cognitive and self-organizing wireless networks has raised the demand to dive deeper into distributed network optimization.

Strictly speaking, optimization is simply the procedure of maximizing some benefit while minimizing a well-defined cost. However, we would like to stress that in wireless networking the term optimization is used rather more freely. Generally, in engineering sciences the aim of optimization is often relaxed to a more realistic design goal, that is to obtain a required performance level while

---

* The authors would like to thank the RWTH Aachen University and the German Research Foundation (Deutsche Forschungsgemeinschaft, DFG) for financial support through the UMIC research center. We also thank the European Union for providing partial funding of this work through the FARAMIR project.

W. Elmenreich, F. Dressler, and V. Loreto (Eds.): IWSOS 2013, LNCS 8221, pp. 103–114, 2014.

satisfying some constraints. Especially in self-organizing wireless networks where decentralized and distributed approaches are called for, the standard definition of an optimization problem by means of a single optimization function that expresses the overall system performance is not practically useful. Instead, a distributed framework is needed, where each network node aims to optimize its own individual utility function.

The general aim of this paper is to introduce the notion of phase transitions into wireless networking optimization. Our initial motivation emerges from the fact that phase transitions are already exploited in optimization theory and have proven to be a very powerful tool in modeling dynamical and statistical physics systems (see, for example, [1–4]). It has been recognized that threshold phenomena, analogous to phase transitions in physical systems, occur in several optimization problems. Actually, there is a considerable and increasing amount of literature that attempts to draw a connecting line between optimization problems and the theory of phase transitions from statistical physics. Despite that, the notion of phase transition has not yet been considered in wireless networking optimization. There is a limited amount of literature addressing phase transitions occurring in wireless networks from a rather theoretical perspective (see, for example, [5–9]), but to the best of our knowledge phase transitions have not been addressed in the context of wireless networking optimization problems.

We argue that the study of phase transitions in wireless networking optimization problems is important not only from a theoretical point of view, but can be also useful in attacking practical optimization problems in a more comprehensive way. A phase transition corresponds to a large, abrupt deviation in the behavior of the problem, and in most cases this change has a significant effect in the performance as well. Taking into account the highly dynamic characteristics of wireless networks we would like then to end up with a solution that lies sufficiently away from a point where a phase transition occurs, so that a small drift from the obtained solution will not cause the system to undergo a phase transition. Therefore, a phase transition can be identified and taken into account in constructing a solution methodology, or even in defining the mathematical representation of the problem in a way that the emergence of undesirable solutions, that lie near a phase transition point, is eliminated.

As a case study we demonstrate a power control optimization problem in a macrocell-femtocell network scenario. More specifically, we suggest that in many cases a hard optimization constraint can be interpreted as a phase transition effect since it defines a threshold that usually corresponds to a qualitative change in the system. We show how such phase transition effects can be mathematically incorporated by means of a dummy variable; this is a typical technique used in econometrics for modeling qualitative changes. In our scenario we apply this approach in formulating a distributed, utility-based representation of the problem in order to prevent the occurrence of poor solutions.

The rest of the paper is organized as follows. In Section 2 we elaborate on the main idea and motivation behind this work. In Section 3 we introduce our example scenario; in Subsection 3.1 we present the system model and the phase transition arising from the standard-form formulation of the optimization problem, and in Subsection 3.2 we discuss the distributed, utility-based approach of the problem. Finally, the paper is concluded in Section 4.

## 2    Motivation for Taking Phase Transitions into Account

The notion of phase transitions originates in physics and has been extensively studied, particularly in the field of statistical mechanics [10]. A system can exist in a number of different phases, which can be distinguished from each other by a set of suitably chosen and well-defined parameters. The macroscopic properties of a system in different phases usually differ significantly. A phase transition is a transition of the system from one phase to another caused by a small change of an external parameter, called control parameter. More specifically, there is a well-defined value of the control parameter, termed critical point, above which one phase exists and as the control parameter goes below that value a new phase appears. Briefly, a phase transition is an *abrupt* and *qualitative* change in the macroscopic properties of the system. One of the most widely known examples is the liquid-gas phase transition, where temperature plays the role of the control parameter under the assumption of constant pressure. A phase transition is also characterized by an order parameter. The order parameter is usually a quantity which is zero in one phase (usually above the critical point), and obtains a non-zero value as the system enters the new phase.

There is an inherent connection between statistical physics and optimization theory. After all, the main task in an optimization problem is to find a configuration of the problem parameters that minimizes a cost function, which is analogous to finding the minimal energy state in statistical physics. Within the context of optimization problems a phase transition is simply an abrupt change in the behavior of the problem. The identification of an existing phase transition contributes to a better understanding of the problem that is not only of theoretical interest, but can be also useful in solving optimization problems more competently.

However, optimization tasks in wireless networking scenarios constitute a special case due to the dynamic characteristics of the wireless environment. Unlike classical optimization problems, a wireless networking optimization task is executed on a wireless network which is not isolated from the outside world, meaning that the system we have to deal with is subject to external influences. Consequently, finding an optimal or a sufficiently good solution is not always enough as in the case of classical optimization of static systems because in a wireless network several factors might easily cause the system to drift from the solution after it has been reached. Under these dynamic conditions not only the optimality of the solution, but also the stability of the performance is an important factor that has to be taken into account. The behavior of the system around a selected solution point needs to be considered as well. As discussed above a phase transition is synonymous to an abrupt change in the behavior; if a phase transition is located near to a selected solution point, then a slight drift from this point might lead to a large undesirable fluctuation in the system performance. Under these circumstances, even the optimal solution might be rather undesirable if it is highly unstable in terms of performance. Therefore, if a phase transition is likely to cause a large fluctuation affecting the performance of the network, this fact has to be taken into account in solving the optimization problem. For instance, the solution can be restricted to lie somewhat away from the phase transition point. A general rule regarding whether a phase transition should be taken into account when solving a problem and how it should be treated cannot be established, but a different decision has to be made each time depending on the behavior of the specific problem under study.

Let us consider, for example, an optimization problem with one or more constraints, among which an inequality constraint. The inequality constraint can be addressed as a phase transition that separates the solution space of the problem into two phases, one that corresponds to the subset of feasible solutions and a second phase that corresponds to the unacceptable solutions. Therefore, solving the problem very close to the critical point, which is defined by the value of the inequality constraint, might result in experiencing fluctuations that can easily bring the system into the unacceptable performance regime. However, we need to stress that the appropriate treatment for this phase transition can only be determined according to problem-specific factors. For example, in several cases falling slightly below the value defined by an inequality constraint corresponds to an analogous performance degradation which brings the system at an operational point not satisfying the system requirements. In such a case we might decide that slight violations of the inequality constraint for short time periods can be tolerated. Naturally, a critical factor that we need to take into account in this decision is how sharp is the fluctuation effect induced by the phase transition, that is how drastic are the behavioral changes we experience due to small deviations of the control parameter around the phase transition point. On the other hand, sometimes the inequality constraint might correspond to a point below which the system collapses. In these cases we should select an approach that will guarantee that the constraint will not be violated, i.e, the solution will lie sufficiently away from the phase transition point.

The phase transition dynamics and analysis we are introducing can be seen as a complementary, or sometimes alternative tool, for the better known game theoretical analysis of Nash equilibrium. Phase transitions in the context of utility optimization provides different mathematical framework from game theory, and in a sense it emphasizes the need to analyze the stability of optimal solutions and consequences of perturbations around those points.

## 3    Introducing the Notion of Phase Transitions in a Power Control Scenario

We demonstrate the main idea of the paper by introducing the notion of phase transitions in a power control optimization problem. More specifically we consider an uplink power control problem in a two-tier femtocell network. This scenario has become really very interesting due to the fact that femtocells of-fer an attractive possibility to increase the capacity of wireless networks. We stress that our aim in this paper is not to present a novel solution methodology for the specific power optimization task, but rather to propose a different and complementary optimization approach – based on the notion of phase transitions – that can be used to address a multitude of other wireless networking optimization problems as well.

## 3.1    The System Model

We consider a macrocell-femtocell scenario where a single Macrocell Base Station (MBS) is underlaid with $M$ Femtocell Access Points (FAPs). The system model is illustrated in Figure 1. The macrocell and the femtocells have access

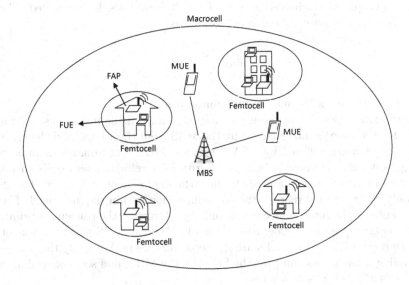

**Fig. 1.** A two-tier femtocell network where a cellular macrocell is underlaid with shorter range femtocells

to the same set of $K$ subchannels, $\{0, 1, ..., k, ..., K - 1\}$. The cellular network operator owning the MBS is the so-called licensed user of the spectrum, whereas the femtocells – which are small, low-power networks deployed in homes and enterprises – are allowed to make use of the same channels (frequencies) under the strict requirement that they will not disrupt the operation of the macrocell, i.e., a specified Signal to Interference-plus-Noise Ratio (SINR) for the macrocell communications needs to be guaranteed anytime. We assume that a subchannel allocation scheme operates independently of power control. After the subchannel assignment a power control scheme can be applied independently in each sub-channel. Therefore, without loosing generality, we shall be concerned with the uplink power control in a single subchannel, $k$. We suppose that a subset of $N$ femtocells, $\{0, 1, ..., n, ..., N-1\}$, are assigned with subchannel $k$. Each femtocell might serve several users, but obviously only one user will use subchannel $k$ in each of the $N$ femtocells. Therefore, we will concentrate only on a single Femtocell User Equipment (FUE) in each femtocell $n$, denoted as $FUE_{n_k}$. We suppose that the subchannel $k$ is currently used also by the MBS in order to serve a Macrocell User Equipment, $MUE_k$. This fact introduces a macrocell-femtocell interference in our scenario, whereas the reuse of subchannel $k$ by more than one femtocells causes a femtocell-to-femtocell interference. Since we concentrate only on subchannel $k$, for the sake of brevity the subscript $k$ shall be omitted in the rest of the paper.

Assuming that the transmission power in the macrocell is already determined, the goal is to find the uplink transmission powers, $\{p_1, p_2, ..., p_n, ..., p_N\}$, of the FUEs in each of the $N$ femtocells so that the total femtocell capacity, that is the sum of the capacities achieved in each of the $N$ femtocells, is maximized. Thus, the objective function can be defined as follows:

$$\max_{p_n} \sum_{i=1}^{N} C_n, \tag{1}$$

where $C_n$ is the capacity achieved in femtocell $n$.

Clearly two types of interference constraints arise in the uplink case; one accounts for the interference suffered by the MBS from the FUEs, and the second for the interference suffered by a FAP from the MUE and from FUEs in neighboring femtocells. However, the first priority of a cellular operator is to fully satisfy all macrocell users, and shall allow the deployment of cochannel femtocells only if they do not disrupt the communication within the macrocell. Thus, we can address the interference constraints by borrowing the concept of primary and secondary users from Cognitive Networks [11]; the MBS plays the role of a primary user, whereas the FAPs are the secondary users. We apply this concept by defining a hard constraint for the SINR at the MBS, and soft constraints for the SINRs of the FAPs as follows

$$\gamma_M \geq \gamma_{M_{target}} \qquad \text{(Hard Constraint)} \tag{2}$$
$$\gamma_{F_n} \geq \gamma_{F_{min}}, \forall n \in \{1, 2, ..., n, ...N\} \quad \text{(Soft Constraint)} \tag{3}$$

In optimization theory a hard constraint is a constraint that must be satisfied. On the other hand, soft constraints are supposed to be followed to the extent that this is possible, but not at the expense of the hard constraints.

Although femtousers will not be necessarily fully satisfied, a soft SINR constraint for every FAP aims to preserve the bit rate in all the femtocells at an acceptable level in order to avoid an extremely unfair capacity assignment among the femtocells. On the other hand the macrouser must remain fully satisfied all the time, meaning that a violation of the hard constraint (Equation 2) is not tolerated. The hard constraint can be translated into a phase transition separating two phases in the solution space of the optimization problem, one corresponding to feasible solutions, and the other to unacceptable solutions. The received SINR at the MBS acts as the control parameter of the phase transition and the target value defined by the corresponding constraint, $\gamma_{M_{target}}$, is the critical value.

Therefore, we need to take care that a solution algorithm for this particular scenario shall not converge to a solution which is too close to the phase transition point. In this case the critical point is precisely known, thus if we plan to solve the problem in a centralized fashion (based on Equations 1, 2 and 3) we can just shift the critical point in order to include a safety margin, $\epsilon$, by slightly modifying the corresponding constraint (Equation 2) as follows

$$\gamma_M \geq \gamma_{M_{target}} + \epsilon. \tag{4}$$

Naturally, the value of the margin, $\epsilon$, has to be determined according to the severeness of the performance fluctuations around the critical point.

Nevertheless, in practice the solution to this issue is usually less trivial. In the following subsection we will be discussing a self-organizing, distributed approach for this scenario.

## 3.2  A Utility-Based Distributed Approach

In the previous subsection we discussed the occurrence of a phase transition in an uplink power control problem for a macrocell-femtocell scenario. The definition of the optimization problem as presented in Section 3.1 corresponds to a centralized approach of the problem and thus is not particularly attractive from a more practical point of view. In this section we shall consider a distributed, self-organizing approach of the problem based on utility functions. The communicating pairs of FAPs and FUEs are responsible to determine their own transmission powers. For this purpose we assume that the elements of each pair exchange control information between them. Specifically, for the case of uplink power control the task is to determine the transmission powers of each FUE and we suppose that the corresponding FAP is informing the FUE about its received SINR.

The utility function of each FUE consists of a reward term, which assigns a payoff increasing with the SINR achieved at the FBS, and a penalty term, which penalizes the interference caused by the FUE to other FAPs of neighboring femtocells, and to the MBS. The reward term corresponds to the objective function of the optimization problem (Equation 1), and the penalty term accounts for the constraints (Equations 2 and 3). Additionally, a variable $\beta$ needs to be introduced in order to control the interaction between the reward term and the penalty. Therefore, we write the utility function for $FUE_n$ as follows

$$U_n = R_n(\gamma_{F_n}, \gamma_{F_{min}}) - \beta P_n(p_n, \gamma_{F_n}). \tag{5}$$

We define the reward function as follows [12]

$$R_n(\gamma_{F_n}, \gamma_{F_{min}}) = 1 - e^{\left(-\alpha(\gamma_{F_n} - \gamma_{F_{min}})\right)} \tag{6}$$

and the penalty term as

$$P_n(p_n, \gamma_{F_n}) = p_n \gamma_{F_n}. \tag{7}$$

The reward is negative if the achieved SINR is below the target SINR, it becomes 0 when the SINR is equal to the target SINR and increases exponentially towards 1. The rate of this exponential increase is controlled by parameter $a$. On the other hand, the penalty term discourages the FUE to increase its transmission power unlimitedly by assigning a high cost if the transmission power and the achieved SINR are high. We shall be considering one of the simplest strategies that can be adopted. The FUEs initialize their transmission powers at some minimum value, $p_{min}$, and continue by increasing the transmission power in small steps if such a decision will increase their utility function.

The quantitative results that shall be presented in the following discussion are obtained by a Matlab simulation of this distributed, utility-based scheme. We consider a MBS and an outdoor MUE separated by a distance equal to 250 meters. Then, we consider that 50 FBSs use channel $k$ within the coverage area of the MBS and are placed at random locations within 300 meters from the MBS. The FUEs are within a radius of 20 meters from their FBS, and the FBSs are at least 40 meters away from each other (i.e., we assume that the coverage areas of FBSs do not overlap with each other). All FUEs are considered to be indoors, in the same building with their associated FBS, but within a different building from neighboring FBSs. The pathloss models we use follow the 3GPP LTE (Long Term Evolution) specification [13]. We also consider lognormal shadowing with standard deviation equal to 4 dB for communication within a femtocell, and 8 dB for every other case [13]. We consider Additive White Gaussian Noise (AWGN) with variance $\sigma^2 = N_o B$, where $N_o$ is the noise power spectral density and $B$ is the subchannel bandwidth. The noise power spectral density is set to $-174$ dBm/Hz and the subchannel bandwidth is 200 KHz.

The solution where the network will converge is determined by the value of the parameter $\beta$. As already discussed, the femtocells should not be deployed at the expense of the macrocell performance, hence a violation of the hard constraint of Equation 2 is not tolerated. Consequently, an appropriate selection of $\beta$ must guarantee the satisfaction of this constraint. Figure 2 illustrates the achieved SINR at the MBS for different values of $\beta$. The different curves correspond to results obtained for different network topologies. Clearly, the outcome is highly variable for different random topologies indicating that the determination of the parameter $\beta$ is definitely not a straightforward task. However, in both cases the SINR increases approximately linearly with $\beta$. Actually, in practice the SINR requirements are imposed by the data rate required to provide the services accessed by the user. From the user's point of view the achieved data rate is the performance. If we plot the achieved capacity based on the Shannon's law instead of the SINR (Figure 3) we observe that the relation with $\beta$ is not linear in this case and for the lower values of $\beta$ the performance in terms of capacity is even more sensitive to small changes of $\beta$. Therefore, there is a region where even small errors in the determination of an appropriate value for $\beta$ can affect

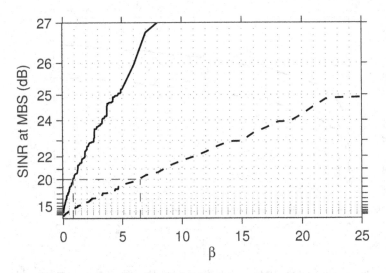

**Fig. 2.** The SINR received at the MBS (linear scale) against the parameter $\beta$ for two different random topologies. The target SINR at the MBS ($\gamma_{M_{target}}$), i.e., the critical point, is set to 20 dB.

drastically the macrocell capacity. The dashed red lines in Figure 3 indicate the values corresponding to the critical SINR value of 20 dB at the MBS.

In utility-based approaches it is a common tactic to account for inequality constraints by introducing penalty terms. Nevertheless, the two representations are not precisely the same; the penalty term cannot define a precise threshold like an inequality constraint. As discussed in Section 3.1, the hard inequality constraint of Equation 2 defines basically the critical point between two phases, one corresponding to the acceptable and the other to unacceptable solutions. Although this phase transition in the solution space is quite straightforward from the standard constraint-based representation of the optimization problem, this effect is not clearly mapped in the distributed approach. More precisely, what is missing from the utility-based approach is the effect of a drastic degradation of the utilities as the SINR at the MBS crosses the critical point, that will model the phase transition from acceptable to unacceptable solutions. Towards this direction, we shall create this phase transition effect by means of a binary variable. The usage of binary (or dummy) variables to model structural changes[1] is commonly employed in econometrics [14]. We define the utility functions as follows

$$U_n = R_n(\gamma_{F_n}, \gamma_{F_{min}}) - (\beta + \delta)P_n(p_n, \gamma_{F_n}). \tag{8}$$

---

[1] In econometrics terminology such phenomena are called structural changes, but they are clearly analogous to what we call phase transitions.

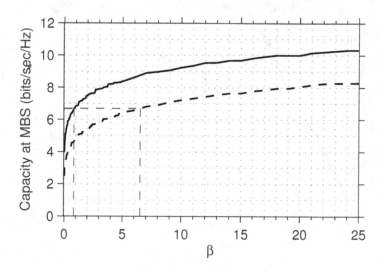

**Fig. 3.** The normalized capacity for the MBS against the parameter $\beta$ for four different network topologies. The dashed red lines indicate the values corresponding to the critical SINR value of 20 dB at the MBS.

The binary variable is the parameter $\delta$, which can be defined as follows

$$\delta = \begin{cases} x, \text{ if } \gamma_M < \gamma_{M_{target}} - \epsilon \\ 0, \text{ otherwise} \end{cases} \tag{9}$$

where $x$ is a very large value which makes the penalty very expensive if the SINR at the MBS is below the critical value, so that a further increase of the transmission powers is prohibited. Initially, the FUEs set the parameter $\delta$ to 0. The MBS shall inform all the FUEs if its received SINR is below the threshold value, and then the FUEs shall change the value of $\delta$ to $x$. The parameter $\epsilon$ is just a safety margin from the critical point. Figure 4 illustrates the SINR at the MBS when using the utility function of Equation 8 for exactly the same configurations as those used to produce Figures 2 and 3. Basically, the binary variable $\delta$ does not allow the system to undergo a phase transition towards the regime of unacceptable solutions.

Finally, we would like to stress that the method of introducing a binary variable in order to model an existing phase transition is not specific to this particular scenario presented here. In general, a binary variable can be used to model a discrete shift, or equivalently, it can distinguish two alternatives. Therefore, the proposed approach of introducing a binary variable in modeling a utility function is in general suitable for treating two phases differently like, for example, in case we want to favor a solution lying in the one side of a phase transition.

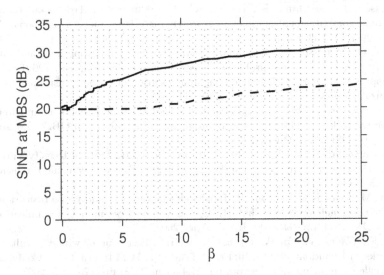

**Fig. 4.** The SINR received at the MBS against the parameter $\beta$ after introducing the binary variable $\delta$ in the utility functions

## 4   Conclusions

In this paper we addressed phase transitions in the context of wireless networking optimization. A phase transition is often translated into an abrupt and drastic change in the behavior of an optimization problem. Especially in wireless networking, such effects can result in a significant performance instability because a network might easily drift from a solution point due to the dynamic environment and conditions. Therefore, we argue that in several cases the awareness of a phase transition should be taken into account when solving the problem in order to ensure the stability of the solution.

As a case study we presented a power control optimization problem for a macrocell-femtocell network. The specific scenario exhibits a phase transition between two regions of the solution space, one corresponding to acceptable and the other to unacceptable solutions. We showed how we can formulate a distributed, utility-based approach of the optimization problem in order to map clearly this effect of distinguishing between a favorable and undesirable phase by means of a binary variable. This approach is suitable if we want to treat two phases in a different way, for instance, if we want to favor solutions lying in the one side of a phase transition.

## References

1. Hartmann, A., Weigt, M.: Phase Transitions in Combinatorial Optimization Problems. Wiley (2005)

2. Monasson, R., Zecchina, R., Kirkpatrick, S., Selman, B., Troyansky, L.: Determining computational complexity from characteristic 'phase transitions'. Nature 400(6740), 133–137 (1999)
3. Martin, O., Monasson, R., Zecchina, R.: Statistical mechanics methods and phase transitions in optimization problems. Theor. Comput. Sci. 265, 3–67 (2001)
4. Achlioptas, D., Naor, A., Peres, Y.: Rigorous location of phase transitions in hard optimization problems. Nature 435, 759–764 (2005)
5. Krishnamachari, B., Wicker, S., Bejar, R.: Phase transition phenomena in wireless ad hoc networks. In: Global Telecommunications Conference, GLOBECOM 2001, vol. 5, pp. 2921–2925. IEEE (2001)
6. Krishnamachari, B., Wicker, S., Bejar, R., Pearlman, M.: Critical Density Thresholds in Distributed Wireless Networks, pp. 279–296. Kluwer Academic Publishers, USA (2003)
7. Durvy, M., Dousse, O., Thiran, P.: Border effects, fairness, and phase transition in large wireless networks. In: The 27th Conference on Computer Communications, INFOCOM 2008, pp. 601–609. IEEE (April 2008)
8. Xing, F., Wang, W.: On the critical phase transition time of wireless multi-hop networks with random failures. In: Proc. of the 14th ACM International Conference on Mobile Computing and Networking, MobiCom 2008, pp. 175–186 (2008)
9. Michalopoulou, M., Mähönen, P.: Game theory for wireless networking: Is a nash equilibrium always a desirable solution? In: 2012 IEEE 23rd International Symposium on Personal Indoor and Mobile Radio Communications (PIMRC), pp. 1249–1255 (September 2012)
10. Reichl, L.: A Modern Course in Statistical Physics. Wiley (1998)
11. Broderson, R., Wolisz, A., Cabric, D., Mishra, S., Willkomm, D.: Corvus: A cognitive radio approach for usage of virtual unlicensed spectrum. White Paper (2004)
12. Chandrasekhar, V., Andrews, J., Muharemovic, T., Shen, Z., Gatherer, A.: Power control in two-tier femtocell networks. IEEE Transactions on Wireless Communications 8(8), 4316–4328 (2009)
13. 3GPP TR36.814: Further advancements for E-UTRA, physical layer aspects. Technical report (March 2010), http://www.3gpp.org/
14. Greene, W.: Econometric Analysis. Prentice Hall (2003)

# On the Local Approximations of Node Centrality in Internet Router-Level Topologies*

Panagiotis Pantazopoulos, Merkourios Karaliopoulos, and Ioannis Stavrakakis

Department of Informatics and Telecommunications
National & Kapodistrian University of Athens, Ilissia, 157 84 Athens, Greece
{ppantaz,mkaralio,ioannis}@di.uoa.gr

**Abstract.** In many networks with distributed operation and self-organization features, acquiring their global topological information is impractical, if feasible at all. Internet protocols drawing on node centrality indices may instead approximate them with their egocentric counterparts, computed out over the nodes' *ego-networks*. Surprisingly, however, in router-level topologies the approximative power of localized *ego-centered measurements* has not been systematically evaluated. More importantly, it is unclear how to practically interpret any positive correlation found between the two centrality metric variants.

The paper addresses both issues using different datasets of ISP network topologies. We first assess how well the *egocentric* metrics approximate the original *sociocentric* ones, determined under perfect network-wide information. To this end we use two measures: their rank-correlation and the overlap in the top-$k$ node lists the two centrality metrics induce. Overall, the rank-correlation is high, in the order of 0.8-0.9, and, intuitively, becomes higher as we relax the ego-network definition to include the ego's $r$-hop neighborhood. On the other hand, the top-$k$ node overlap is low, suggesting that the high rank-correlation is mainly due to nodes of lower rank. We then let the node centrality metrics drive elementary network operations, such as local search strategies. Our results suggest that, even under high rank-correlation, the locally-determined metrics can hardly be effective aliases for the global ones. The implication for protocol designers is that rank-correlation is a poor indicator for the approximability of centrality metrics.

## 1 Introduction

Computer networks are generally complex systems, typically owing their complexity to their topological structure and its dynamic evolution but also the heterogeneity of their nodes. *Social Network Analysis* (SNA) [21] has been lately viewed as a solid analytical framework for understanding their structural properties and the interactions of their elements. The expectation within the networking community is that socio-analytical insights can benefit the analysis of network structure and the design of efficient protocols. Indeed, there has been evidence that such insights can improve network functions such

* This work has been partially supported by the National & Kapodistrian University of Athens Special Account of Research Grants no 10812 and the European Commission under the Network of Excellence in Internet Science project EINS (FP7-ICT-288021).

W. Elmenreich, F. Dressler, and V. Loreto (Eds.): IWSOS 2013, LNCS 8221, pp. 115–126, 2014.

as content-caching strategies in wired networks [5] and routing/forwarding in opportunistic networks [6].

Common denominator to these efforts is the use of SNA-driven metrics for assessing the relative centrality (*i.e.*, importance) of individual network nodes, whether humans or servers. The computation of these metrics, however, typically demands *global* information about all network nodes and their interconnections. The distribution and maintenance of this information is problematic in large-scale networks. In self-organized environments lacking centralized management operations, in particular, it may not even be an option at all. A more realistic alternative for assessing node centrality draws on its ego network, *i.e.*, the subgraph involving itself, its 1-hop neighbors, and their interconnections. *Egocentric* measurements, carried out within their immediate locality, let nodes derive *local approximations* of their centrality.

Lending to simpler computations, egocentric[1] metrics have, in fact, found their way into protocol implementations [6], [5]. Nevertheless, the capacity of these local approximations to substitute the globally computed sociocentric metrics is almost always taken for granted rather than evaluated. Over Internet router-level topologies, in particular, a systematic study of the approximative power of egocentric metrics is missing.

In this paper, we focus on node centrality metrics, which are the most commonly used in networking protocols. Besides the well known Betweenness Centrality (BC) metric, we consider the Conditional BC (CBC), the destination-aware BC variant proposed in [16], which is particularly suited to many-to-one communication and data flow typologies. We first question how well do sociocentric metrics, computed under global topological information, *correlate* with their egocentric variants, as computed locally over the nodes' ego networks. Since in most protocol implementations, it is the *ranking* of the metric values that matters rather than their *absolute values*, we measure this correlation over the node rankings they produce, via their full *rank*-correlation and the overlap in the top-$k$ node sets in the two rankings. Then, contrary to previous works, we proceed to study what the measured correlation coefficients can reveal regarding the capacity of rank-preserving local centrality metrics to substitute the original global metrics in protocol primitives.

*Related work*: Computing centrality values typically requires global topological information. BC computations, in particular, can be carried out by exact yet efficient [3] or approximation algorithms [4]. The latter seek to provide accurate sampling methods over the considered network node pairs. Some recently proposed random walk schemes that reveal high BC [11] or central nodes in a general sense [10], provide *distributed*-fashion approaches but require gathering beyond-local-scope information. The efforts to devise *local* approximations of the global centrality metrics span various disciplines and usually study *whether* do the former correlate with the latter. From a networking standpoint, local approximations of centrality metric have been proposed for the bridging [13] and closeness centrality [22]. Both works report high positive rank-correlation scores between the two counterparts over synthetic and real world networks, but do not explore whether this is sufficient to benefit networking operations. Betweenness Centrality, on the other hand, is locally approximated over real-world network topologies with the node degree, or Degree Centrality (DC), due to the evidenced linear correlation between the

---

[1] The terms egocentric and local as well as sociocentric and global are used interchangeably.

mean BC and DC metrics [20]. However, this holds for AS-level Internet maps that can exhibit different connectivity properties compared to the router-level topologies; or, in the case of *load*, a BC variant, for scale-free networks [9]. Finally, some social studies provide experimental evidence for positive correlation between sociocentric and ego-centric BC over *small social* and *synthetic* networks [12], [7].

*Our contribution*: We report on the BC *vs*. egoBC, BC *vs*. DC, CBC *vs*. egoCBC *correlation* over synthetic topologies and then focus on real-world router-level topologies of sizes up to 80K nodes, coming from 20 different ISPs (Section 3). Our synthetic networks yield significant rank-correlation that weakens as their size increases. In router-level topologies, in almost all cases, we measure high rank correlation (0.8-0.9), which becomes even higher when we compute the egocentric betweenness variants over generalized ego networks, corresponding to their 2-hop neighborhoods. In each case, we analyze the time complexity and message overhead related to the computation of the metrics. (Section 2).

On the contrary, the top-$k$ overlap is found low prompting us to study whether the rank-preserving local metrics can *in practice* substitute the global ones. Our experiments with basic network functions such as the centrality-driven content search, show (Section 4) that high rank-correlation values between the two BC variants are poor predictors for the capacity of the local approximations to substitute the original global metrics in actual network protocol operations.

## 2   Sociocentric *vs.*Egocentric Centrality Metrics

The sociocentric metrics of betweenness centrality (BC) and the conditional betweenness centrality (CBC), along with their egocentric counterparts[2] are first presented. Then, the complexity savings achieved by the latter, are discussed.

### 2.1   Globally Computed Centrality Indices

Consider an arbitrary node pair $(s, t)$ over a connected undirected graph $G = (V, E)$. If $\sigma_{st}$ is the number of shortest paths between $s$ and $t$ and $\sigma_{st}(u)$ those of them passing through node $u \neq t$, then the betweenness centrality of node $u$, equals $BC(u) = \sum_{\substack{s,t \in V \\ s<t}} \frac{\sigma_{st}(u)}{\sigma_{st}}$. Effectively, $BC(u)$ assesses the importance of a network node for serving information that flows over shortest paths in the network [8]. Whereas $BC(u)$ is an average over all network node pairs, the conditional betweenness centrality index (CBC), captures the topological centrality of node $u \neq t$ with respect to a *specific* destination node $t$ [17] and is given by: $CBC(u; t) = \sum_{\substack{s \in V \\ s \neq t}} \frac{\sigma_{st}(u)}{\sigma_{st}}$, with $\sigma_{st}(s) = 0$. Therefore, CBC is particularly suited to settings where information is directed towards a particular node with discrete network functionality.

### 2.2   Locally Computed Centrality Metrics - Ego Networks

Computing the above metrics requires information about the whole network topology and implies computational and message load overheads. In distributed settings, where

---

[2] The degree centrality (DC) is by default an egocentric centrality metric.

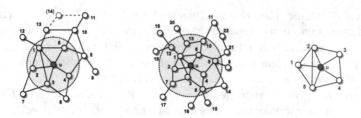

a. Ego network of node $u$ ($r = 1$).   b. Ego network of node $u$ ($r = 2$)   c. Computing $egoBC(u; 1)$

**Fig. 1.** a) Nodes 2, 3 and 4 contribute to $egoCBC(u; 11, 1) = 2$ with contributions 1/2, 1/2 and 1, respectively. b) Node 8 reaches the exit node 10 for destination node 11 through five different paths, two of which pass through node $u$, thus contributing 2/5 to $egoCBC(u; 11, 2)$. c) Toy-example computation: $egoBC(u; 1) = 10(1 - \frac{6}{10}) - [2(1 - \frac{1}{2}) + 2(1 - \frac{1}{3})] = \frac{5}{3}$.

nodes may be energy constrained or no explicit centralized network management be available, these computations are not favorable or not an option at all. Instead, techniques for *locally* assessing the centrality of network nodes can be borrowed from the SNA concepts. In the so-called *ego-network* structure of social studies the person we are interested in is referred to as the "ego" and its ego-network comprises itself together with those having an affiliation or friendship with it, known as "alters". Alters may as well share relations with each other (Fig. 1.a).

Hereafter, we generalize the ego-network definition to include nodes (alters) lying $r$ hops away from $u$ and the edges (links) between them. Formally, we can define the $r^{th}$-order ego network as follows. Let $N_r^u$ be the set of nodes that form the $r$-hop neighborhood around $u$, i.e., $N_r^u = \{n \in G : 1 \leq h(n, u) \leq r\}$, where $h(a, b)$ denotes the minimum hopcount between nodes $a$ and $b$. The $r^{th}$-order ego network of node $u$ is the graph $G_r^u = (V_r^u, E_r^u)$, where the set of nodes and edges are $V_r^u = N_r^u \cup \{u\}$ and $E_r^u = \{(i, j) \in E : i, j \in V_r^u\}$, respectively. For $r = 1$ the network $G_1^u$ corresponds to the original ego network definition and consists of $|V_1^u| = DC(u) + 1$ nodes and $|E_1^u| = DC(u) + CC(u) \cdot \binom{DC(u)}{2}$ edges, where $CC(u)$ is the clustering coefficient [14] of node $u$ and $DC(u)$ its degree centrality. Practically, values of $r > 2$ would tend to cancel the advantages that local ego-centered measurements induce.

Accordingly, BC metrics of a certain node can be defined with respect to its ego network. For the egocentrically measured betweenness centrality ($egoBC$) of node $u$, it suffices to apply the typical BC formula over the graph $G_r^u$ : $egoBC(u; r) = BC(u)_{|V=V_r^u}$. We further detail the computation formula of $egoBC(u, 1)$. In Fig. 1.c the ego-node $u$ is connected to $DC(u)=5$ first-neighbor nodes which are partially interconnected with each other. The maximum egoBC value that $u$ can attain, when no links between its neighbors are present, equals $\binom{5}{2}$. From this value we need to subtract the number of pairs (like nodes 2-5 or 3-4) that share a direct link and therefore communicate without traversing $u$; their number equals the numerator of the clustering coefficient $CC(u)$. Finally, we need to carefully account for those node pairs that share no direct link but are connected with multiple 2-hop paths, such as nodes 2 and 4 that are connected via the paths 2-3-4, 2-u-4 and 2-5-4. The idea is that only one of those paths will cross $u$; thus, we need to discount the original contribution (i.e., unit) of each non-directly connected node pair $i, j \in N_1^u$ by as much as the inverse of the number of competing 2-hop paths connecting $i$ and $j$. This number corresponds to the element $(i, j)$ of matrix $A^2$, where $A$ is the adjacency matrix for $G_1^u$. Summing up, we have:

$$egoBC(u; 1) = \begin{cases} \binom{DC(u)}{2}(1 - CC(u)) - \sum_{i,j \in N_1^u : h(i,j)=2}(1 - \frac{1}{A_{i,j}^2}) & \text{if } DC(u) > 1 \\ 0 & \text{if } DC(u) = 1 \end{cases} \quad (1)$$

The egocentric counterpart of conditional betweenness centrality ($egoCBC$), on the other hand, is less straightforward. For each ego network and for a given destination node $t$, we need to identify the set of *exit nodes* $e_r(u; t) = \{t' \in N_r^u : h(u, t') + h(t', t) = h(u, t)\}$, i.e., all nodes $r$ hops away from the ego node $u$ that lie on the shortest path(s) from $u$ to $t$. This set is effectively the *projection* of the remote node $t$ on the local ego network and may be a singleton but never the null set. In Fig. 1.a, for example, we have $e_1(u; 11) = \{6\}$, $e_1(u; 9) = \{4, 6\}$ for the $G_1^u$ while in Fig. 1.b we have $e_2(u; 11) = \{10\}$, $e_2(u; 14) = \{5, 8\}$ for the $G_2^u$. For each node $s \in G_r^u$, we need to calculate the fraction of shortest paths from $s$ towards *any* of the nodes in $e_r(u; t)$ that traverse the ego node. Thus the egocentric variant of CBC is given by

$$egoCBC(u; t, r) = \sum_{\substack{s \in V_r^u \\ t' \in e_r(u;t)}} \frac{\sigma_{st'}(u)}{\sigma_{st'}} 1_{\{h(s,t') \leq h(s,l), \, l \in e_r(u;t)\}} \quad (2)$$

Again, in Fig. 1a, node 4 contributes to the $egoCBC(u, 11)$ value since its shortest path to the single exit node 6 traverses $u$, although it has a shorter path to node 11, via nodes $\{5,10\}$ that lie outside the ego network. Likewise, its contribution is a full unit, rather than $1/2$, since the second shortest path to node 6 passes through node 5, a node outside the ego network of $u$. This is the price egocentric metrics pay for being agnostic of the world outside their $r$-neighborhood. Although, the definitions of both ego- and sociocentric metrics are valid under weighted and unweighted graphs, we focus on the latter ones. The way link weights affect the correlation operations is clearly worth of a separate study.

### 2.3 Complexity Comparison of betweenness Counterparts

We briefly discuss how the two types of metrics compare in terms of message overhead and time complexity required for their computation (see Table 1). Message overhead is measured in messages times the number of edges they have to travel. In both cases, we can distinguish two metric computation phases: the collection of topological information and the execution of computations.

Sociocentric computation of centrality. The network nodes need to collect global information about the overall network topology; hence, each one of the $|V|$ network nodes has to inform the other $|V| - 1$ about its neighbors. This generally requires $O(|E_f|)$ message copies and $O(D)$ time steps for each node's message, where $D$ is the network diameter and $|E_f|$ the number of edges in the flooding subgraph. In the best case, the flooding takes place over the nodes' spanning trees, hence the message overhead is $O(|V| - 1)$. For the distribution of one round of messages by all nodes, the overhead becomes $O(|V|^2)$; the time remains $O(D)$ assuming that the process evolves in parallel. With knowledge of the global topology, each node can compute the BC values of all other nodes in the network. An efficient way to do this is to invoke Brandes' algorithm [3], featuring $O(|V| \cdot |E|)$ complexity for unweighted graphs and $O((|E| + |V|) \cdot |V|log|V|)$ complexity for weighted graphs. Interestingly, the CBC values of each network node with respect to all other network nodes, emerge as intermediate results of Brandes' algorithm for the BC computation [3].

Egocentric computation of centrality. Intuitively, the egocentric variants save complexity. The message overhead over the whole network is $O(2 \cdot |E|)$ for the ego network with $r = 1$ and $O(2 \cdot d_{max}|E|)$ for the ego network with $r = 2$, where $d_{max}$ is the maximum node degree; for dense graphs this overhead becomes $O(|V|^2)$. The time required for the distribution of information is of no concern, $O(1)$. The egoBC and egoCBC computation for $r = 1$ can be carried out as in [7]. The computation involves a multiplication of an $O(d_{max})$-size square matrix and trivial condition checks. For $r = 2$, we can employ [3] replacing $|V|$ with $d_{max}^2$. Finally, DC is considered to be immediately available to every node.

**Table 1.** Complexity comparison of socio-vs.ego-centric metrics

| Metric | Time complexity | Message overhead |
|---|---|---|
| BC | $O(|V|^3)$ | $O(D \cdot |V|)$ |
| egoBC (r=1) | $O(d_{max}^3)$ | $O(2 \cdot |E|)$ |
| egoBC (r=2) | $O(d_{max}^4)$ | $O(2 \cdot d_{max} \cdot |E|)$ |
| CBC | $O(|V|^3)$ | $O(D \cdot |V|)$ |
| egoCBC(r=1) | $O(d_{max}^3)$ | $O(2 \cdot |E|)$ |
| egoCBC(r=2) | $O(d_{max}^4)$ | $O(2 \cdot d_{max} \cdot |E|)$ |
| DC | $O(1)$ | – |

**Table 2.** Correlation study between BC and egoBC on grid networks

| Grid size | Diameter / Mean degree | Spearman $\rho$ | |
|---|---|---|---|
| | | ego-network (r=1) | ego-network (r=2) |
| 5x5 | 8 / 3.200 | 0.9195 | 0.9679 |
| 10x10 | 18 / 3.600 | 0.8400 | 0.9556 |
| 20x20 | 38 / 3.800 | 0.6802 | 0.8459 |
| 50x50 | 98 / 3.920 | 0.2429 | 0.2942 |
| 60x8 | 66 / 3.717 | 0.5735 | 0.6336 |
| 90x8 | 96 / 3.728 | 0.5390 | 0.5870 |
| 150x8 | 156 / 3.737 | 0.4584 | 0.4181 |
| 400x8 | 406 / 3.745 | 0.1633 | 0.2213 |

As expected, since $d_{max}$ is typically much smaller than $|V|$, the use of local metrics bears apparent computational benefits. The question of whether these metrics correlate well with the sociocentric ones is considered next.

## 3 Experimental Correlation-Operations Study between Socio- and Egocentric Centrality Metrics

### 3.1 Correlation Coefficients and Network Topologies

In comparing the ego- with sociocentric metrics, we are concerned with (rank) correlation-operations. Protocol implementations that utilize highly central nodes, usually care more about the way the metric *ranks* the network nodes rather than their absolute metric values; this is how a degree-based search scheme explores unstructured P2P networks [2] or egoBC is used in DTN forwarding [6]. Alternatively, we may need to employ only a subset of $k$ nodes with the top centrality values [17] and therefore, we

---

[3] The algorithm in [3] effectively visits successively each node $u \in V$ and runs augmented versions of shortest path algorithms. By the end of each run, the algorithm has computed the $|V| - 1$ $CBC(v; u)$ values, $v \in V$; while the $|V|$ $BC(v)$ values result from iteratively summing these values as the algorithm visits all network nodes $u \in V$.

also study the ovelap scores between the top-$k$ nodes determined with ego- and socio-centric counterparts, respectively. We capture the rank correlation in the non-parametric Spearman measure $\rho$, which assesses how monotonic is the relationship between the ranks of the two centrality variables. For the sake of completeness we also measure the well-known Pearson correlation (see [18] for the relevant formulas). The $r_{Prs}$ coefficient also assesses a straight-line relationship between the two variables but now the calculation is based on the actual data values. Both $\rho$ and $r_{Prs}$ lie in [-1,1] and mark a significant correlation between the considered metrics when they are close to 1. For the target-node-dependent $CBC$ values, we present the correlation averages along with the 95% confidence intervals estimated over an at least 7% of the total locations, sample. Regarding the networks we experiment with, our emphasis is on the real-world intradomain Internet topologies. Nevertheless, synthetic graphs with distinct structural properties such as the rectangular grid have been also employed to provide insights. The router-level ISP topologies do not have the predictable structure of the synthetic ones and typically size *up to* a few thousand nodes. Yet, over these topologies networking protocols designed to cope with self-organization requirements, will seek to utilize local centrality metrics [17]. We have experimented with three sets of router-level topologies:

*mrinfo topologies:* The dataset we consider includes topology data from 850 distinct snapshots of 14 different AS topologies, corresponding to Tier-1, Transit and Stub ISPs [15]. The data were collected daily during the period 2004-08 with the help of a multicast discovering tool called `mrinfo`, which circumvents the complexity and inaccuracy of more conventional measurement tools. Herein we experiment with a representative subset of Tier-1 and Transit topologies.

*Rocketfuel topologies:* The Rocketfuel technique [19] has been shown to collect high-fidelity router-level maps of ISPs and therefore has been widely used despite its old, around 2002, publication. The considered dataset includes measurements from 800 vantage points serving as `traceroute` sources. Innovative techniques such as BGP directed probing and IP identifiers, have been applied to reduce the number of probes, and discover the different interface IP addresses that belong to the same router (*i.e.*, alias resolution), respectively.

*CAIDA topologies:* The most recent of our datasets [1] was collected during Oct-Nov 2011 by CAIDA performing `traceroute` probes to randomly-chosen destinations from 54 monitors worldwide. Parsing the provided separate file that heuristically assigns an AS to each node found, we have determined the router-to-AS ownership and subsequently have extracted out of the raw data files the topologies of the nodes operated by certain ASes. Our effort was to discover the largest ISP networks in the dataset. With all three datasets we avail a rich experimentation basis of a diverse Internet topologies' set that can minimize the effect of measurement technique errors.

## 3.2  Experimental Results

We choose to spend more effort experimenting on the BC-egoBC, BC-DC correlation debates, expected to attract more interest (see Section 4) than the limited-scope of CBC.

*egoBC vs.BC* : In grid topologies, ego networks have fixed size depending on their position, *i.e.*, corner, side, or internal nodes. Thus, the egoBC index may only exhibit a small number of different values (*e.g.*, 1, 3 and 6 with respect to the node's location

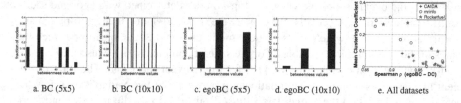

a. BC (5x5)    b. BC (10x10)    c. egoBC (5x5)    d. egoBC (10x10)    e. All datasets

**Fig. 2.** a)-d) Probability distribution of BC and egoBC values for scaling size of a grid network. e) egoBC-DC correlation scales linearly with <CC> values.

when $r = 1$). When grid dimensions grow larger, the number of shortest paths between any node pair grows exponentially, resulting in a richer spectrum of BC values over the grid nodes. On the other hand, the possible egoBC values remain the same; only the distribution of grid nodes over these values changes (see Fig. 2a-d). Consequently, in Table 2, $\rho$ values decrease monotonically with the grid size. Same behavior and reasoning also holds for the line network.

Our findings for the real-world ISP topologies are listed in Table 3. Even with measurements within the first-order ego network, there is high positive rank correlation between BC and egoBC. On the other hand, the Pearson coefficient suggests looser yet positive association. When egoBC is computed in the second-order ego network, both correlation coefficients become even higher and also more similar with each other. The structural characteristics of the considered ISP topologies differ from the grid topologies; their diameter and clustering coefficients attain many different values as their size scales. Provably there is enough asymmetry throughout the topology to yield a wide range of BC and egoBC values that favors high correlation. A notable exception is the mrinfo Level-3 ISP topology (IDs 12,13) which we comprehensively study (and visualize) in [18]. Herein, we provide only a brief explanation. Datasets 12,13 were found to avail some clustered structures of nodes that, interestingly, exhibit higher egoBC than global BC values; it is these nodes that distort the desired linear relation between the centrality counterparts. Finally, note that Table 3 results imply that the localized bridging centrality [13], which uses egoBC to approximate BC, is also highly correlated with its global counterpart in these topologies.

_DC vs.BC_: In grid topologies we have observed the same correlation degradation with the network size, explained along the aforementioned BC-spectrum arguments. In all router-level topologies (Table 3) we find high Pearson and even higher Spearman correlation although consistently lower than the corresponding egoBC _vs._BC one, at least for the Spearman $\rho$. As such, the previously reported [20] DC-BC correlation over AS-level topologies is extended, by our results, to the router-level ones. Finally, the earlier observed egoBC-BC rank correlation can be further justified on the grounds of the high DC-BC Spearman values; the router-level topologies exhibit vanishing clustering coefficients and thus, the egoBC metric attains similar values to DC (see Eq. 1). This is depicted in Fig. 2.e where the egoBC-DC correlation scales linearly with the mean clustering coefficients, especially in the CAIDA set.

_egoCBC vs.CBC_ : We now assess the ego variant of the CBC metric ($r$=1). Table 4 suggests significant positive rank correlation in all studied ISP topologies, even for

**Table 3.** Correlation study between BC-egoBC and BC-DC on router-level ISP topologies

| DataSet ID | | ISP(AS number) | $<\bar{C}C>$ | Diameter | Size | $<$degree$>$ | BC vs.ego-BC | | | | BC vs.DC | |
|---|---|---|---|---|---|---|---|---|---|---|---|---|
| | | | | | | | Spearman $\rho$ | | Pearson $r_{Prs}$ | | Spearman $\rho$ | Pearson $r_{Prs}$ |
| | | | | | | | ego-net. r=1 | ego-net. r=2 | ego-net. r=1 | ego-net. r=2 | | |
| | 35 | Global Crossing(3549) | 0.479 | 9 | 100 | 3.78 | 0.9690 | 0.9853 | 0.7029 | 0.9255 | 0.8506 | 0.6714 |
| m | 33 | NTTC-Gin(2914) | 0.307 | 11 | 180 | 3.53 | 0.9209 | 0.9565 | 0.7479 | 0.8561 | 0.8180 | 0.6664 |
| r | 13 | Level-3(3356) | 0.169 | 25 | 378 | 4.49 | 0.2708 | 0.9393 | -0.0918 | 0.7982 | 0.1953 | -0.0813 |
| i | 12 | -//- | 0.149 | 28 | 436 | 4.98 | 0.2055 | 0.9381 | -0.1217 | 0.7392 | 0.1696 | -0.1128 |
| n | 20 | Sprint(1239) | 0.287 | 16 | 528 | 3.13 | 0.9866 | 0.9928 | 0.5805 | 0.8488 | 0.8543 | 0.6815 |
| f | 38 | Iunet(1267) | 0.231 | 12 | 645 | 3.75 | 0.8790 | 0.9516 | 0.9094 | 0.9568 | 0.8549 | 0.7708 |
| o | 44 | Telecom Italia(3269) | 0.037 | 13 | 995 | 3.65 | 0.7950 | 0.9828 | 0.3362 | 0.8699 | 0.7733 | 0.4852 |
| | 50 | TeleDanmark(3292) | 0.058 | 15 | 1240 | 3.06 | 0.9569 | 0.9738 | 0.5475 | 0.9025 | 0.9388 | 0.5538 |
| R | 61 | Ebone(1755) | 0.115 | 13 | 295 | 3.68 | 0.9736 | 0.9860 | 0.6856 | 0.8895 | 0.9443 | 0.7457 |
| O | 62 | Tiscali(3257) | 0.028 | 14 | 411 | 3.18 | 0.9522 | 0.9659 | 0.6073 | 0.9281 | 0.9464 | 0.7103 |
| C | 63 | Exodus(3967) | 0.273 | 14 | 353 | 4.65 | 0.9125 | 0.9792 | 0.6100 | 0.9061 | 0.8204 | 0.6241 |
| K | 64 | Telstra (1221) | 0.015 | 15 | 2515 | 2.42 | 0.9990 | 0.9990 | 0.3336 | 0.7565 | 0.9783 | 0.5172 |
| E | 65 | Sprint(1239) | 0.022 | 13 | 7303 | 2.71 | 0.9980 | 0.9990 | 0.4770 | 0.7977 | 0.9562 | 0.6537 |
| T | 66 | Level-3(3356) | 0.097 | 10 | 1620 | 8.32 | 0.9841 | 0.9923 | 0.6346 | 0.9075 | 0.9655 | 0.7045 |
| F | 67 | AT&T(7018) | 0.005 | 14 | 9418 | 2.48 | 0.9988 | 0.9994 | 0.3388 | 0.5302 | 0.9882 | 0.4483 |
| L | 68 | Verio (2914) | 0.071 | 15 | 4607 | 3.28 | 0.9904 | 0.9969 | 0.4729 | 0.8044 | 0.9315 | 0.6718 |
| | 70 | UUNet (701) | 0.012 | 15 | 18281 | 2.77 | 0.9841 | 0.9886 | 0.5430 | 0.8752 | 0.9694 | 0.7544 |
| C | 71 | COGENT/PSI(174) | 0.062 | 32 | 14413 | 3.09 | 0.9638 | 0.9599 | 0.7272 | 0.9354 | 0.8940 | 0.8791 |
| A | 72 | LDComNet(15557) | 0.021 | 40 | 6598 | 2.47 | 0.9674 | 0.9245 | 0.3782 | 0.7676 | 0.9479 | 0.6634 |
| I | 74 | ChinaTelecom(4134) | 0.083 | 19 | 81121 | 3.97 | 0.8324 | 0.8986 | 0.7861 | 0.9714 | 0.7370 | 0.8795 |
| D | 75 | FUSE-NET(6181) | 0.018 | 10 | 1831 | 2.38 | 0.9903 | 0.9763 | 0.6205 | 0.8574 | 0.9536 | 0.7445 |
| A | 76 | JanetUK(786) | 0.031 | 24 | 2259 | 2.26 | 0.9819 | 0.9834 | 0.4444 | 0.8506 | 0.9450 | 0.5765 |

the outlier case of the `mrinfo` Level-3 networks. Intuitively, the correlation between CBC counterparts is expected to be higher than the one between BC counterparts; by neglecting the world outside of the ego network, the egoBC inaccuracies (compared to the globally determined BC) may arise anywhere across the network. On the contrary, $egoCBC(u;t,r)$ considers only the paths that lead to the target $t$, somehow focusing on an angle that encompasses $t$; thus, it may differ from the CBC(u;t) view only across that certain angle.

*top-k nodes overlap* : Finally, we assess the extent to which the nodes with the top-$k$ values of local centrality metrics coincide with the corresponding ones determined by the global BC. Interestingly, Table 5 shows that this overlap is low, at least for the first-order ego network measurements. This does not actually contradict our previous results since correlation is determined over all network nodes rather than a subset of cardinality $k$. Clearly, the observed high rank-correlation is mainly due to nodes of lower rank; for instance, those with already zero values for both centrality counterparts (*i.e.*, DC=1) have been reported to drastically contribute to the high egoBC-BC correlation [7]. Along this thread, our comprehensive study [18] has shown that the actual association between the two metric variants is not determined solely by the degree distribution. Finally, the low top-$k$ overlap scores serve as a warning sign as to what the high coefficients can reveal about the practical implications of local centrality metrics.

**Table 4.** Rank-correlation between CBC and ego-CBC (r=1)

| ID | 35 | 33 | 13 | 12 | 20 | 44 |
|---|---|---|---|---|---|---|
| Spearman $\rho$ | 0.9489 | 0.9554 | 0.7336 | 0.7035 | 0.9847 | 0.9902 |
| Conf. Interv. | 0.013 | 0.003 | 0.007 | 0.005 | 0.003 | 0.001 |

| ID | 50 | 61 | 62 | 63 | 75 | 76 |
|---|---|---|---|---|---|---|
| Spearman $\rho$ | 0.9739 | 0.8423 | 0.9321 | 0.7641 | 0.9961 | 0.9853 |
| Conf. Interv. | 0.009 | 0.027 | 0.016 | 0.023 | 0.005 | 0.002 |

**Table 5.** Overlap(%) between nodes with the top-$k$ local centrality and BC values

| ID | k=10 | | | k=30 | | |
|---|---|---|---|---|---|---|
| | egoBC(r=1) | egoBC(r=2) | DC | egoBC(r=1) | egoBC(r=2) | DC |
| 50 | 30.0 | 70.0 | 30.0 | 10.0 | 60.0 | 10.0 |
| 63 | 10.0 | 60.0 | 10.0 | 0.0 | 30.0 | 0.0 |
| 67 | 0.0 | 10.0 | 0.0 | 0.0 | 30.0 | 0.0 |
| 70 | 0.0 | 90.0 | 0.0 | 36.7 | 76.7 | 43.3 |
| 71 | 40.0 | 90.0 | 40.0 | 56.7 | 80.0 | 60.0 |
| 72 | 40.0 | 50.0 | 40.0 | 50.0 | 60.0 | 50.0 |

## 4   Practical Utility of Local Centrality Metrics

We now study basic network operations where local centrality metrics are employed, encouraged by the high corresponding correlation, to substitute the global ones. Our aim is to assess whether such options can actually be effective in networking practice.

*A local-centrality-driven navigation scheme*: In unstructured power-law P2P networks, it has been shown [2] that a high-degree (file) seeking strategy is more efficient than a random-walk strategy. The former, at each step, passes a single query message to a neighboring node of higher degree exploiting the great number of links pointing to high degree nodes. We have implemented a similar navigation scheme that crawls the

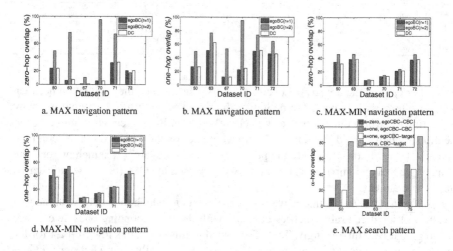

**Fig. 3.** a-e) Overlap between the final locations achieved with local and global centrality as driver; in e) the two rightmost bars of each dataset show the egoCBC/CBC-driven search hit-rates

network following a MAX or MAX-MIN pattern with respect to node centrality; each time, the crawler moves to the neighbor with the maximum(minimum) centrality out of those that exhibit higher(lower) values than the current node, utilizing a self-avoiding path. We randomly select 20% of the network nodes as starting points and execute 10 runs (*i.e.*, crawlings) for each one but with no destinations predetermined. Effectively, we seek to compare the navigation patterns and final locations achieved by using the different centrality variants as drivers. $\alpha$-hops overlap measures the percentage of the final locations lying within $\alpha$ hops away from those the global metric yields. Zero-hop overlap refers to the destinations' exact matching. Regarding the final locations, figs. 3a-d) show that the local metrics ($r=1$) can hardly be effective aliases for the global ones. The crawler, driven by local metrics, is measured [18] to consistently take from 0.4 up to 2.3 on average less hops than when it is BC-driven, failing to identify the same navigation pattern *i.e.*, sequence of central nodes. In view of the high correlation between the involved metrics, we have tried to shed some light on the somewhat counterintuitive, poor navigation performance [18]. Removing sequentially four nodes from a toy topology of initially perfect rank correlation (*i.e.*, $\rho = 1$) between the two counterparts, we measured it reducing to 0.9953. At the same time the *zero*-hop overlap for the MAX pattern drastically diminished from 100% to 61.90%. Clearly, the numerical summaries the coefficients provide, fail to capture in micro-level the relative significance of each node which determines the scheme's performance.

A *local-centrality-driven search scheme*: As the conditional centrality metrics involve a target node, we are enabled to compare CBC and egoCBC essentially over a (content) search scheme. For each starting location we randomly select a target node and seek to reach it utilizing a MAX search pattern. Fig. 3.e shows low ovelap between the final locations achieved by the two counterparts while the hopcount to the final location is again measured consistently lower (*i.e.*, 0.3 to 1.5 hops) for the egoCBC case; driven by the local variant the search fails to track closely the route that the global one induces. In terms of the *one*-hop overlap between the achieved final locations and the targets (*i.e.*, hit-rates), which is also reported in [2], figure 3.e (two rightmost bars) shows that the egoCBC-driven search indeed hits significantly less targets than the CBC-driven does. The number of targets reached by the local CBC variant is in good agreement with the discovered P2P nodes in [2], using DC as a driver; there, a networkwide path of the query may cumulatively hit up to 50% of either the targets or their first neighbors. On the contrary, we obtain high overlap values when the global metric drives the MAX search pattern, since the closer node $u$ lies to the target $t$ the higher its $CBC(u; t)$.

## 5   Conclusions

The paper has questioned to what extent the original centrality metrics can be substituted by their computationally friendly local approximations in router-level topologies. First, the metrics are shown to exhibit high rank-correlation with their local counterparts across all datasets (20 ISPs) but one. On the other hand, the match between the two variants is much worse when we compare the top-$k$ nodes selected by each of them. Then, we tried to assess what the algebraic values of the correlation coefficients reveal regarding the performance of network functions, when the original metrics are substituted by their local approximations. Both a simple navigation and a search scheme employing local centrality metrics produce significantly different navigation patterns and lower hit-rates, respectively, than their counterparts with the original global metrics. These results suggest that, despite the positive correlations, local variants can hardly offer effective approximations to the original metrics. Our work essentially warns against relying on the correlation indices for assessing the substitutability of ego- and sociocentered variants of centrality metrics.

## References

[1] The CAIDA UCSD Macroscopic Internet Topology Data Kit-[ITDK 2011-10], http://www.caida.org/data/active/internet-topology-data-kit/

[2] Adamic, L.A., et al.: Search in power-law networks. Physical Review E 64(4) (September 2001)

[3] Brandes, U.: A faster algorithm for betweenness centrality. Journal of Mathematical Sociology 25, 163–177 (2001)

[4]  Brandes, U., Pich, C.: Centrality Estimation in Large Networks. Int'l Journal of Birfucation and Chaos 17(7), 2303–2318 (2007)
[5]  Chai, W.K., He, D., Psaras, I., Pavlou, G.: Cache "less for more" in information-centric networks. In: Bestak, R., Kencl, L., Li, L.E., Widmer, J., Yin, H. (eds.) NETWORKING 2012, Part I. LNCS, vol. 7289, pp. 27–40. Springer, Heidelberg (2012)
[6]  Daly, E.M., Haahr, M.: Social network analysis for information flow in disconnected delay-tolerant manets. IEEE Trans. Mob. Comput. 8(5), 606–621 (2009)
[7]  Everett, M., Borgatti, S.P.: Ego network betweenness. Social Networks 27(1), 31–38 (2005)
[8]  Freeman, L.C.: A set of measures of centrality based on betweenness. Sociometry 40(1), 35–41 (1977)
[9]  Goh, K.I., et al.: Universal behavior of load distribution in scale-free networks. Phys. Rev. Lett. 87(27) (December 2001)
[10] Kermarrec, A.M., et al.: Second order centrality: Distributed assessment of nodes criticity in complex networks. Comp. Com. 34 (2011)
[11] Lim, Y., et al.: Online estimating the $k$ central nodes of a network. In: IEEE Network Science Workshop (NSW 2011), pp. 118–122 (June 2011)
[12] Marsden, P.: Egocentric and sociocentric measures of network centrality. Social Networks 24(4), 407–422 (2002)
[13] Nanda, S., Kotz, D.: Localized bridging centrality for distributed network analysis. In: IEEE ICCCN 2008, Virgin Islands (August 2008)
[14] Newman, M.E.J.: The Structure and Function of Complex Networks. SIAM Review 45(2), 167–256 (2003)
[15] Pansiot, J.J., et al.: Extracting intra-domain topology from mrinfo probing. In: Proc. PAM, Zurich, Switzerland (April 2010)
[16] Pantazopoulos, P., et al.: Efficient social-aware content placement for opportunistic networks. In: IFIP/IEEE WONS, Slovenia (2010)
[17] Pantazopoulos, P., et al.: Centrality-driven scalable service migration. In: 23rd Int'l Teletraffic Congress (ITC 2011), San Francisco, USA (2011)
[18] Pantazopoulos, P., et al.: On the local approximations of node centrality in Internet router-level topologies. Tech. rep. (January 2013), http://anr.di.uoa.gr/index.php/publications
[19] Spring, N.T., et al.: Measuring ISP topologies with rocketfuel. IEEE/ACM Trans. Netw. 12(1), 2–16 (2004)
[20] Vázquez, A., et al.: Large-scale topological and dynamical properties of the Internet. Phys. Rev. E 65(6), 066130 (2002)
[21] Wasserman, S., Faust, K.: Social network analysis. Cambridge Univ. Pr. (1994)
[22] Wehmuth, K., Ziviani, A.: Distributed assessment of the closeness centrality ranking in complex networks. In: SIMPLEX, NY, USA (2012)

# Modelling Critical Node Attacks in MANETs

Dongsheng Zhang[1] and James P.G. Sterbenz[1,2]

[1] Information and Telecommunication Technology Center
Department of Electrical Engineering and Computer Science
The University of Kansas, Lawrence, KS 66045, USA
{dzhang,jpgs}@ittc.ku.edu
http://www.ittc.ku.edu/resilinets
[2] School of Computing and Communications, InfoLab21
Lancaster University, Lancaster LA1 4YW, UK
jpgs@comp.lancs.ac.uk

**Abstract.** MANETs (mobile ad hoc networks) operate in a self-organised and decentralised way. Attacks against nodes that are highly relied to relay traffic could result in a wide range of service outage. A comprehensive model that could enhance the understanding of network behaviour under attacks is important to the design and construction of resilient self-organising networks. Previously, we modelled MANETs as an aggregation of time-varying graphs into a static weighted graph, in which the weights represent link availability of pairwise nodes. Centrality metrics were used to measure node significance but might not always be optimal. In this paper, we define a new metric called *criticality*[1] that can capture node significance more accurately than centrality metrics. We demonstrate that attacks based on criticality have greater impact on network performance than centrality-based attacks in real-time MANETs.

**Keywords:** graph theory, MANET, network resilience, challenge modelling, centrality, criticality.

## 1 Introduction and Motivation

MANETs have the merit of quick and flexible self-organisation and have been utilised in various scenarios, such as vehicular ad hoc networks and wireless sensor networks. With the increasing deployment of MANETs in commercial and military uses, it becomes vital to design and construct a resilient and survivable MANET. Because of the peer-to-peer and multi-hop properties of MANET communications, challenges against certain nodes might cause the partitioning of the network. By strengthening specific critical nodes such as increasing the transmission range or recharging the battery, the whole network could be more resilient under attacks and challenges. Furthermore, due to node mobility, unpredictably long delay, and channel fading of wireless environment [19], MANETs suffer from dynamic connectivity that increases the complexity of modelling.

---

[1] Our definition is different from the critically *k*-connected graph defined in [12].

W. Elmenreich, F. Dressler, and V. Loreto (Eds.): IWSOS 2013, LNCS 8221, pp. 127–138, 2014.

In our previous work, MANETs are modelled as time-varying graphs and pairwise node interactions are aggregated within specific time windows [21]. Dynamic MANETs can be represented as static weighted graphs, in which the weight refers to link availability. The adversary is assumed to have complete knowledge about the network. Centrality metrics can be used to identify the relative significance of each node. However, research in SNA (social network analysis) showed that the removal of high centrality nodes might not necessarily cause maximal loss of network connectivity [4]. Articulation points whose removal increases the number of connected components could lead to maximum degradation of overall network performance. The CNPs (critical node problems), which are generally defined as the detection of a subset of nodes whose removal disconnects the graph maximally, have been widely studied in SNA [2]. Instead of using weighted centrality metrics to indicate node significance, we provide a more accurate selection of nodes whose removal could have a higher impact on the network. Network simulations are performed to verify how attacks against nodes selected by this approach could impact overall network performance.

The rest of the paper is arranged as follows. In Section 2, we introduce background and related work about wireless network challenge modelling, centrality metrics, and CNPs. In Section 3, we illustrate the difference between critical node detection in weighted and unweighted graphs. We describe how to detect critical nodes in weighted graph and model malicious attacks based on node criticality in Section 4. In Section 5, we exploit simulations to verify our approach using several examples with plots showing network performance under various types of attacks. Finally, we summarise our work and mention the steps for future research in Section 6.

## 2    Background and Related Work

Understanding network challenges that are inherent in the self-organising networks is essential to construct a resilient and survivable network [18]. Simulation tools can be utilised to examine network performance under various attacks and challenges [14]. Centrality metrics can be used to measure relative node significance. However, those nodes whose removal could partition the topology might be more vital to the whole network.

### 2.1    Network Challenge Modelling

A simulation framework that evaluates realistic challenges in wired networks has been developed [6]. Due to the dynamics and channel properties of wireless networks, techniques used to improve the disruption tolerance and network reliability for wired networks are not enough in the wireless context [19]. In order to capture the time-varying characteristics of MANETs, temporal graph metrics used in SNA take into account topology evolutions over time [20]. However, they are not applicable to real-time MANETs since traditional MANET routing protocols do not allow data transmission if there is no route between source and destination at the time of sending, which makes metrics such as temporal path ineffective. Temporal network robustness is used to measure how communication of a given time-varying network is affected by random attacks [16]; however, it does not address the impact of critical node attacks that could result in higher degradation of network performance.

## 2.2  Centrality Metrics

Centrality metrics (degree, betweenness, and closeness) have been used to measure comparative node importance in both weighted and unweighted graphs in SNA [9,15]. Degree centrality indicates the node communication ability within its neighbours and the disadvantage is that it only captures the relation between adjacent nodes and fails to take into account global topological properties. Metrics that can capture global properties include betweenness and closeness. Betweenness is defined as the frequency that a node falls on the shortest paths and closeness is defined as the inverse of the sum of the shortest paths [9]. In order to calculate node betweenness and closeness in a weighted graph, the weights need to be inverted to represent link cost instead of link availability [13]. Betweenness measures the degree to which a node enables communication between all node pairs. Closeness measures the extent to which node communications capabilities are independent of other nodes. However, they might not always be effective to definitively indicate the structural importance of each node, since those nodes whose removal could cause most damage on the network are not necessarily the nodes with high centrality values [4]. Examples will be presented in Section 3 to illustrate the difference.

## 2.3  Critical Node Problems

Vulnerability assessment in cases of potential malicious attacks is critical to network resilience design. A framework that models the network as a connected directed graph can evaluate network vulnerability by investigating how many nodes are required to be removed so that network connectivity can be degraded to a desired level [8]. A general graph-theoretical formulation of this problem is removing a certain number of nodes in a graph to maximize the impairment against overall network connectivity, which falls under CNPs. The CNPs are known as $\mathcal{NP}$-hard on general graphs [2]. Heuristics, branch and cut algorithms, and dynamic programming algorithms have been proposed to solve CNPs; however, all of them put certain constraints on graph structures such as trees, series-parallel graphs, or sparse graphs [2,7,17]. As far as we know, no effective approximation algorithms for the weighted graph CNPs have been proposed. Critical node behaviour has been studied using network simulations by only considering discrete static connected topologies [10,11] and cannot be extended to general self-organising and dynamic MANETs.

## 3   Critical Nodes in Unweighted and Weighted Graphs

CNPs deal with the detection of one or multiple nodes whose removal would result in minimal pairwise connectivity. Two examples are given to show the relationship between nodes with high centrality values and the most critical nodes in unweighted and weighted graphs.

Figure 1a presents a 10-node unweighted graph topology, in which node 4 has the highest degree, betweenness, and closeness centrality values. However, the deletion of node 7 instead of node 4 from the graph would partition the network and result in lower connectivity within the rest of the graph. In contrast, the network is still connected after deleting node 4. The impact of removing node 4 is of no significant difference to the removal of any other nodes except node 7.

The case is more complex for critical nodes detection in a weighted graph. Figure 1b has the same network topology as Figure 1a except that each link is associated with certain weight. Assuming that the weights of link $(4,7)$ and $(6,7)$ are 0.001, whereas weights of all other links are far higher than 0.001. The removal of node 7 has trivial impact on the entire network, since node 7 is weakly connected to node 4 and 6 compared to other links before being removed. Even though the removal of node 7 partitions the network, the significance of the articulation point for an unweighted graph cannot be directly extended to weighted graphs. In the next section, we propose a method that can handle critical nodes detection approximately in a weighted graph.

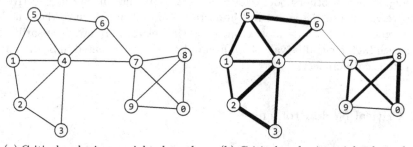

(a) Critical nodes in unweighted graph    (b) Critical nodes in weighted graph

**Fig. 1.** Critical node problems for unweighted and weighted graphs

## 4   Modelling Approach

In our modelling, two wireless nodes are assumed to be adjacent if they are within the transmission range of each other (without interference) and are connected if they can be reached via multihop links. Symmetric communication between nodes is assumed and an undirected graph is sufficient to model the network. Previously, we have modelled the dynamics of the MANETs by aggregating network topologies into a weighted graph in which the weight represents link availability, that is, the fraction of time that nodes are adjacent given the dynamic self-organisation of the network [21]. Based on this weighted link availability model, we propose a new method to detect critical nodes.

## 4.1   Constructing Link Availability Graphs

Figure 2 presents a scenario of MANET topologies at six consecutive time steps and Figure 3 shows the aggregated representation as an adjacency matrix. In Table 4.1, three centrality metrics are calculated based on the weighted adjacency matrix and they give different indications of node significance. Node 5 has the highest degree of 2.66; node 4 has highest betweenness of 0.4; node 3 has the highest closeness of 0.409. We define *criticality* as the inverse of the sum of pairwise path availability after the removal of certain number of nodes. In this case, if only one node will be attacked, the removal of node 5 will impact the network most heavily with the rest of the graph having the minimum connectivity. The detailed algorithm for calculating node criticality is described in Section 4.2.

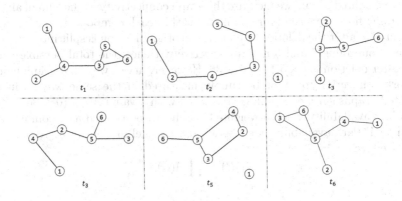

**Fig. 2.** MANET topologies at six consecutive time steps

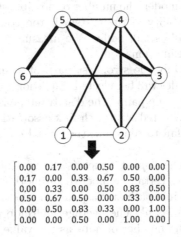

$$\begin{bmatrix} 0.00 & 0.17 & 0.00 & 0.50 & 0.00 & 0.00 \\ 0.17 & 0.00 & 0.33 & 0.67 & 0.50 & 0.00 \\ 0.00 & 0.33 & 0.00 & 0.50 & 0.83 & 0.50 \\ 0.50 & 0.67 & 0.50 & 0.00 & 0.33 & 0.00 \\ 0.00 & 0.50 & 0.83 & 0.33 & 0.00 & 1.00 \\ 0.00 & 0.00 & 0.50 & 0.00 & 1.00 & 0.00 \end{bmatrix}$$

**Fig. 3.** Weighted link availability graph and adjacency matrix

**Table 1.** Single node attack priorities based on centrality and criticality values

| Node | Degree | Betweenness | Closeness | Criticality |
|------|--------|-------------|-----------|-------------|
| 1 | 0.67 | 0.0 | 0.243 | 0.082 |
| 2 | 1.67 | 0.0 | 0.384 | 0.097 |
| 3 | 2.16 | 0.2 | **0.409** | 0.111 |
| 4 | 2.00 | **0.4** | 0.399 | 0.112 |
| 5 | **2.66** | 0.1 | 0.408 | **0.138** |
| 6 | 1.50 | 0.0 | 0.313 | 0.108 |

## 4.2    Measuring Network Connectivity of Weighted Graphs

In an unweighted graph, we measure the graph connectivity as the sum of all possible traffic flows between pairwise nodes within each component. For example, in Figure 1a, after the deletion of node 7, the original graph is split into two components containing 6 and 3 nodes respectively. Hence, the total possible traffic flows after deletion is $6 \times 5 + 3 \times 2 = 36$. However, in our weighted graph model, network connectivity cannot be simply measured in the same way as in unweighted graphs since each link is associated with a value $A(l_i)$, $(0 \leq A(l_i) \leq 1)$ as its link availability. Such a weighted graph can be treated as a complex system model. Path availability for a series and parallel model can be respectively calculated as:

$$A_s(P) = \prod A(l_i) \tag{1}$$

$$A_p(P) = 1 - \prod [1 - A(l_i)] \tag{2}$$

where path $P = (l_1, l_2, ..., l_i)$ [3]. For a general weighted graph that cannot be reduced to a series-parallel model, the number of possible states is non-polynomial and the numerical availability of the system is too complex to compute even after deleting each possible set of nodes. We measure network connectivity by applying following approximations.

**Approximation 1:** Only the strongest path of all possible paths (if there exists one) for a pair of nodes will be selected. Equation 1 can be used to calculate path availability between node pairs. The disadvantage is that if the aggregated graph is close to fully-connected with each link associating with similar weight, the selected path might fail to represent the actual connectivity between node pairs.

**Approximation 2:** The *hopcut*, which is the maximum number of hops in considered paths, is set to a certain number to shrink the size of candidate paths, as the number of all possible paths for a pair of node in a graph of order $n$ ($n$ vertices) could be as high as $n!$. The average multihop path availability tends to decrease with a growing number of hops as the value of $A(l_i)$ is less than 1. Hopcut can be set approximately according to the diameter of the graph. Simulation results will be given in next section about how the hopcut approximation affects the accuracy of critical nodes detection.

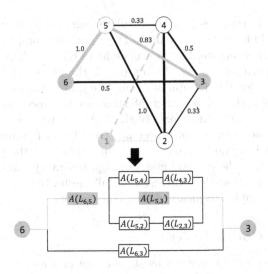

**Fig. 4.** Path selection between node 3 and 6

---

**Algorithm 1.** Overall network availability of a weighted graph $G = (V, E_w)$

---

$A_{\text{sum}} = 0$ {initialise the sum of path availability for all node pairs}
**for** $s$ in $V$ **do**
  **for** $d$ in $V$ **do**
    **if** $s != d$ **then**
      $A_{\text{max}} = 0$ {initialise maximum path availability between a specific node pair}

      **for** $P$ in all paths within hop count $h$ between node $s$ and $d$ **do**
        $A_P = 1$ {initialise current path availability}
        **for** $e$ in $P$ **do**
          $A_P = A_P \times W(e)$ {multiply availability of all the links in the path}
        **end for**
        **if** $A_P > A_{\text{max}}$ **then**
          $A_{\text{max}} = A_P$ {select the path with highest availability}
        **end if**
      **end for**
    **end if**
    $A_{\text{sum}} = A_{\text{sum}} + A_{\text{max}}$ {add up path availability for each pair of nodes}
  **end for**
**end for**
**return** $A_{\text{sum}}$

---

An example that illustrates the algorithm is shown in Figure 4. Node 1 and all the links incident to it are removed due to attack. We need to calculate path availability for pairwise nodes. Consider node pair 3 and 6. The hopcut is set as 3. Hence, all the paths (less than 4 hops) between node 6 and 3 are $\{6,5,3\}, \{6,5,4,3\}, \{6,5,2,3\}$, and $\{6,3\}$ with $A(l_{6,5}) \times A(l_{5,3})$ yielding a highest path availability as $1.0 \times 0.83 = 0.83$. Repeat the same process for other node pairs. Algorithm 1 describes how to calculate pairwise path availability for a general weighted link availability model. The algorithm complexity depends on the weighted graph structure. For an $n$-node MANET with hopcut set as $h$, the complexity of Algorithm 1 ranges from $O(n^2)$ if the aggregated graph is a tree to $O(n^2 h!)$ if the aggregated graph is a complete graph. For the weighted graph model that we use in the paper, the graph structure becomes more full-mesh-like with larger aggregation window sizes of the MANET topologies. However, the difference among node significance becomes trivial in a full-mesh-like network and attacks toward any node would have similar impact on the network. Hence, in order to study how the removal nodes of high centrality and criticality could impact the network, we are less interested in full-mesh-like graphs that are more computationally complex to measure their connectivity.

### 4.3   Detecting Critical Nodes

Instead of using centrality metrics as significance indicators, we directly look into the most critical nodes in the network. It is known that finding critical nodes in a general graph is $\mathcal{NP}$-hard [2]. We propose two approximations that are specific to our link availability model to simplify the algorithm.

**Approximation 1:** Due to the mobility of MANETs and frequent changes of routing tables, it takes a certain amount of time to populate updated routing information on all nodes. Too short a contact duration between nodes might not allow the traffic transmission. Therefore we can simplify the weighted model by deleting links that are lower than a certain threshold.

**Approximation 2:** Since centrality metrics are related to the relative significance of each node, instead of considering all nodes as potential candidates of critical nodes, we only consider those nodes with high centrality values. The main purpose of this paper is to present a simulation model for challenges against MANETs. Even though we apply the above approximations, the computational complexity for graphs with a large number of nodes is still high. In our simulation, the number of nodes is set as 20 and the maximum critical node set size is set as 8. We will not provide a rigorous proof of how close this approximation is to the optimal solution, but we have simulation results in the next section to show that the removal of critical nodes detected using our approach does have higher impact on the whole network. The procedures of detecting $k$-critical nodes of a weighted graph $G$ of order $n$ are presented as follows:

**Step 1.** *Calculate the centrality metrics (degree, betweenness, and closeness) for each node in the graph and store the sorted node list in D, B, C*

**Step 2.** *Let L be the critical node list and add those nodes with k highest centrality values into L*

$$L = D(0:k) \cup B(0:k) \cup C(0:k) \tag{3}$$

*where $D(0:k)$ denotes the first k elements in list D. Remove the duplicate nodes in L.*

**Step 3.** *Let the size of list L be S, then there are $\binom{S}{k}$ different potential critical node sets. Let N be one of the potential critical node set, and remove nodes in N and all the edges that are incident to them from graph G. Calculate the pairwise link availability for the rest of graph as A(G). Iterate the same process on each case in $\binom{S}{k}$.*

**Step 4.** *The set of critical nodes whose removal result in the lowest A(G) is selected as the critical node set.*

## 5   Simulation Analysis

In this section, we employ network simulation to examine the impact of attacks based on different metrics on the network performance. We use the network simulator ns-3.16 as our simulation tool [1]. Constant bit rate UDP traffic is generated at steady state during the simulation and all simulations are averaged over 10 runs. The Gauss-Markov mobility model is used to represent node motion patterns [5]. Functional verification of how different parameters such as node velocity, number of nodes, and routing protocols could influence base scenario network performance without attacks was done in our previous work [21]. In the real world, the placement of network resources must be balanced to the optimised resilience and deployment [18]. Due to space constraints, the simulation parameters used in this paper are limited to 20 nodes, 6 neighbour count, and [5, 10] m/s node velocity range. We examine the impact of a different number of simultaneous node failures, hopcut used in path availability approximation, and windows size that determines the granularity of topology aggregation. For the centrality-based attacks, each metric will be recalculated adaptively after the removal of other nodes. AODV (ad hoc on-demand distance vector) and DSDV (destination-sequenced distance vector) routing protocols are used so that we can have both reactive and proactive routing protocols. PDR (packet delivery ratio) is used to measure the network performance under attacks.

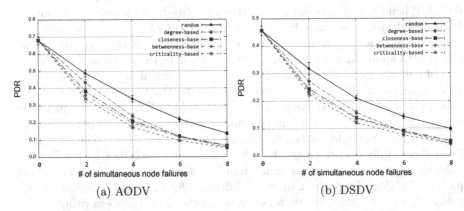

(a) AODV            (b) DSDV

**Fig. 5.** Network performance with increasing number of node failures

In Figure 5, network performance with increasing number of simultaneous node failures arising from random or malicious attacks is examined. All centrality- and criticality-based attacks have apparently higher impact on the network performance than random node failures as expected. Generally, criticality-based attacks result in a lower bound of network performance than centrality-based attacks due to the inaccuracy of node significance predicted by centrality values for certain topologies. With the increase of the number of simultaneous node failures, the difference between centrality- and criticality-based attacks shrinks. This can be explained as when the number of node failures increases, the network is partitioned into multiple components of small order and the difference of node significance become minor. In addition, the degree metric has a more accurate indication of node significance for a relatively large number of simultaneous node failures, whereas betweenness and closeness predict node significance better with a small number of simultaneous node failures.

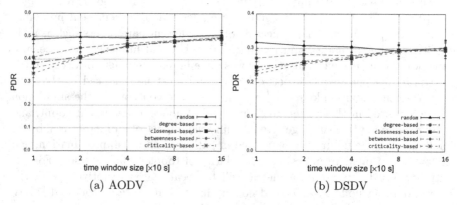

(a) AODV                      (b) DSDV

**Fig. 6.** Network performance with increasing window sizes

Figure 6 shows the influence of topology aggregation granularity. As the time window increases, the difference of relative significance between nodes becomes trivial. When time window size is 10 s, the criticality-based attack has the lowest PDR and degree-based attack has the highest PDR of all centrality based attacks. When time window size is 160 s, PDRs under all different type of attacks converge to approximately the same value. This can be explained as the distribution of pairwise node interactions becomes even among all nodes and nodes will have similar centrality and criticality values under the Gauss-Markov mobility model given a long enough time window, that is, MANET routing is constantly re-optimising the network with moving nodes. As we can see, both centrality and criticality metrics might not be able capture relative node significance accurately in an almost fully-connected graph with evenly assigned weights. Figure 7 shows different approximations of maximum number of hops considered in calculating path availability. There are almost no difference for 3, 4, and 5 hopcut, which means that most communications between pairwise nodes happen within 3 hops for this specific simulation scenario.

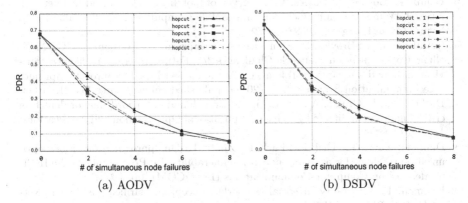

(a) AODV                    (b) DSDV

**Fig. 7.** Network performance with increasing hopcut

# 6  Conclusion and Future Work

In this paper, we conducted a detailed examination of how centrality metrics can be used to indicate node significance in MANETs. We proposed a novel *criticality* approach to measure connectivity of weighted graphs with the weight ranging from 0 to 1. We provided an approximate algorithm to find the critical node subset in MANETs. We demonstrated that critical node attacks impact network performance more than attacks based on centrality values. Future work will include more accurate scalable heuristics to detect any number of critical nodes in weighted graphs of larger size. The impact of network size and density on the detection of critical nodes will also be studied.

**Acknowledgments.** We would like to acknowledge Egemen K. Çetinkaya for the discussions and comments about this paper and other members of the ResiliNets group for discussions on this work. This research was supported in part by NSF FIND (Future Internet Design) Program under grant CNS-0626918 (Postmodern Internet Architecture) and by NSF grant CNS-1050226 (Multilayer Network Resilience Analysis and Experimentation on GENI).

# References

1. The ns-3 network simulator (July 2009), http://www.nsnam.org
2. Arulselvan, A., Commander, C.W., Elefteriadou, L., Pardalos, P.M.: Detecting critical nodes in sparse graphs. Computers and Operations Research 36(7), 2193–2200 (2009)
3. Billinton, R., Allan, R.: Reliability Evaluation of Engineering Systems. Plenum Press London (1983)
4. Borgatti, S.P.: Identifying sets of key players in a social network. Comput. Math. Organ. Theory 12(1), 21–34 (2006)

5. Camp, T., Boleng, J., Davies, V.: A survey of mobility models for ad hoc network research. Wireless Communications and Mobile Computing 2(5), 483–502 (2002), http://dx.doi.org/10.1002/wcm.72

6. Çetinkaya, E.K., Broyles, D., Dandekar, A., Srinivasan, S., Sterbenz, J.P.G.: Modelling Communication Network Challenges for Future Internet Resilience, Survivability, and Disruption Tolerance: A Simulation-Based Approach. Springer Telecommunication Systems, 1–16 (2011) (published online: September 21, 2011)

7. Di Summa, M., Grosso, A., Locatelli, M.: Branch and cut algorithms for detecting critical nodes in undirected graphs. Computational Optimization and Applications, 1–32 (2012)

8. Dinh, T., Xuan, Y., Thai, M., Park, E., Znati, T.: On approximation of new optimization methods for assessing network vulnerability. In: Proceedings of the IEEE Conference on Computer Communications (INFOCOM), pp. 1–9. IEEE (2010)

9. Freeman, L.: Centrality in social networks conceptual clarification. Social Networks 1(3), 215–239 (1979)

10. Kim, T.H., Tipper, D., Krishnamurthy, P., Swindlehurst, A.: Improving the topological resilience of mobile ad hoc networks. In: 7th International Workshop on Design of Reliable Communication Networks (DRCN), pp. 191–197 (October 2009)

11. Kim, T., Tipper, D., Krishnamurthy, P.: Connectivity and critical point behavior in mobile ad hoc and sensor networks. In: IEEE Symposium on Computers and Communications, ISCC, pp. 153–158. IEEE (2009)

12. Kriesell, M.: Minimal connectivity. In: Beineke, L.W., Wilson, R.J. (eds.) Topics in Structural Graph Theory, pp. 71–99. Cambridge University Press (2012)

13. Newman, M.E.J.: Scientific collaboration networks. ii. shortest paths, weighted networks, and centrality. Phys. Rev. E 64(1), 16132 (2001)

14. Nicol, D.M., Sanders, W.H., Trivedi, K.S.: Model-based evaluation: From dependability to security. IEEE Transactions on Dependable and Secure Computing 01(1), 48–65 (2004)

15. Opsahl, T., Agneessens, F., Skvoretz, J.: Node centrality in weighted networks: Generalizing degree and shortest paths. Social Networks 32(3), 245–251 (2010)

16. Scellato, S., Leontiadis, I., Mascolo, C., Basu, P., Zafer, M.: Evaluating temporal robustness of mobile networks. IEEE Transactions on Mobile Computing 12(1), 105–117 (2013)

17. Shen, S., Smith, J.C.: Polynomial-time algorithms for solving a class of critical node problems on trees and series-parallel graphs. Networks 60(2), 103–119 (2012)

18. Sterbenz, J.P.G., Hutchison, D., Çetinkaya, E.K., Jabbar, A., Rohrer, J.P., Schöller, M., Smith, P.: Resilience and survivability in communication networks: Strategies, principles, and survey of disciplines. Computer Networks 54(8), 1245–1265 (2010)

19. Sterbenz, J.P.G., Krishnan, R., Hain, R.R., Jackson, A.W., Levin, D., Ramanathan, R., Zao, J.: Survivable mobile wireless networks: issues, challenges, and research directions. In: Proceedings of the 3rd ACM workshop on Wireless Security (WiSE), Atlanta, GA, pp. 31–40 (2002)

20. Tang, J., Musolesi, M., Mascolo, C., Latora, V.: Temporal distance metrics for social network analysis. In: Proceedings of the 2nd ACM Workshop on Online Social Networks, pp. 31–36 (2009)

21. Zhang, D., Gogi, S.A., Broyles, D.S., Çetinkaya, E.K., Sterbenz, J.P.: Modelling Attacks and Challenges to Wireless Networks. In: Proceedings of the 4th IEEE/IFIP International Workshop on Reliable Networks Design and Modeling (RNDM), pp. 806–812. St. Petersburg (October 2012)

# Evolution as a Tool to Design Self-organizing Systems

István Fehérvári and Wilfried Elmenreich

Institute for Networked and Embedded Systems,
Alpen-Adria-Universität Klagenfurt
{forename.surname}@aau.at

**Abstract.** Self-organizing Systems exhibit numerous advantages such as robustness, adaptivity and scalability, and thus provide a solution for the increasing complexity we face within technical systems. While they are attractive solutions, due to their nature, designing self-organizing systems is not a straightforward task. Artificial evolution has been proposed as a possible way to build self-organizing systems, but there are still many open questions on how an engineer should apply this method for this purpose. In this paper we propose a design methodology for evolving self-organizing systems, that marks the major cornerstones and decisions the designer has to face, thus providing a practical set of guidelines.

**Keywords:** self-organizing systems, evolutionary design, emergence, complexity.

## 1 Introduction

Self-organizing systems are inherently scalable, adaptive and tend to be very robust against single-point failures, which made them an appealing choice in the technical domain [1]. Typically, a design procedure of such systems would be to find the local, micro-level interactions that will result in the desired global behavior. Unfortunately, there is no straightforward way for the design of these rules yet so that the overall system will show the desired macro-level properties.

There exist different approaches to design self-organizing systems. Gershenson [2] introduces a notion of "friction" between two components as a utility to design the overall system iteratively. Auer et al. [3] deduce the local behavior by analyzing a hypothetical omniscient "perfect" agent. The usefulness of evolutionary algorithms to evolve cooperative behavior is demonstrated by Quinn et al. [4] by evolving teamwork strategies among a set of robots. The examples are not limited to robotic applications; Arteconi [5] applies evolutionary methods for designing self-organizing cooperation in peer-to-peer networks. A generic, universal approach to tackle the aforementioned design issues is to use some kind of an automated search, namely evolution to find the right set of local rules that will *drive* the system into the desired macro behavior.

In the domain of mathematical optimization or artificial intelligence metaheuristics, evolutionary methods have been already proven to be especially useful and so are still widely applied. Emergence and evolution have been studied by many scientists from different disciplines, like biology [6], mathematics [7], physics [8], social systems [9], and economic systems [10]. However, most work describes mechanisms, but does not give answers as to how to use those mechanisms to achieve an intended technical effect.

W. Elmenreich, F. Dressler, and V. Loreto (Eds.): IWSOS 2013, LNCS 8221, pp. 139–144, 2014.

With respect to this question, the work by Belle and Ackley on evolvable code [11] and the work on learning for multi-agent systems [12] by Shoham and Leyton-Brown are of special interest for the research goals of the proposed project.

The idea of using this technique to design self-organizing systems has already been mentioned [13], however a deeper, step-by-step guide tailored for this domain is still missing. In this paper, we are presenting a practical set of guidelines in order the fill this gap and support engineers and scientist using evolutionary design.

In the following section, we give reference to related work on design of self-organizing systems with a special focus on evolutionary design. The basic idea of using evolution as a design tool is introduced in Section 2. In Section 3, we present a design methodology for evolving self-organizing systems that marks the major aspects of the approach. In the following subsection, we discuss each aspect of our design approach and introduce guidelines for practical application. The robot soccer case study described in Section 4 shows the application of the mentioned principles in practice. The paper is concluded in Section 5.

## 2  Design by Evolution

Typically, problems based on bio-inspired principles such as self-organization call for bio-inspired solutions. An approach would be to cope with the high complexity is to use some kind of guided automated search, namely evolutionary computation.

The main idea is to create an appropriate system model with its environment, where the components are represented in such a way, that their individual behaviors (i.e. set of local rules) can be iteratively improved via a heuristic algorithm. In other words, the components have to be *evolvable*.

Although the concept of evolutionary design is fairly simple, when it comes to application, it starts to become increasingly complex. In general, the whole process can be decomposed into three major steps: modeling the system and its environment; an iterative search, that explores new solutions; and a final validation phase (see Figure 1).

From an engineering point of view, the modeling phase is the upmost priority, where the crucial decisions have to be made in order to obtain useful results. Among many things this includes selecting the right computation model, the way of interactions and the whole evolutionary model (genotype to phenotype mapping, search algorithm, etc.). The next step is to let the whole system run on its own, thus gradually developing a viable solution with no or very little user interaction. The last phase for any design process is to verify the obtained results. Usually, this means a set of white or black box testing, but in the case of complex networked systems, novel approaches are necessary. Fortunately, there is a lot of ongoing research on how to evaluate such systems [14].

## 3  Design Methodology

As mentioned in the previous section, building a proper system model is essential for a successful evolutionary design. In this paper we propose a methodology for the evolutionary design of self-organizing systems, that identifies the major decisions and

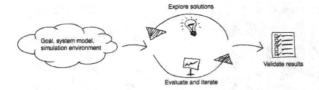

**Fig. 1.** Basic flowchart of the evolutionary design

considerations that the designer must face during the modeling process. As can be seen in Figure 2, we distinguish 6 major components:

- The task description
- The simulation setup
- The interaction interface
- The evolvable decision unit
- The search algorithm
- The objective function

Typically, every engineering problem starts with a task description that properly identifies the objectives and constraints in detail. In other words, this can be seen as a contract that describes the expectations of the desired system on a high abstraction level. Consequently, the task description has a high influence on the applied simulation model and naturally on the objective function.

The next step is then to specify a system model based on the task description that is not only efficient, but accurately represents the important aspects of the system to be modeled while abstracting unnecessary factors. However, this step should not include how the components are represented or the way they interact with each other, since these would involve several further important decisions. For this purpose we separate them into an interaction interface unit and to an evolvable decision unit. In the former, one should plan the way the components of the system can interact with each other and their environment, thus not only defining the communication possibilities (i.e. sensors), but also their underlying interfaces (i.e. protocols).

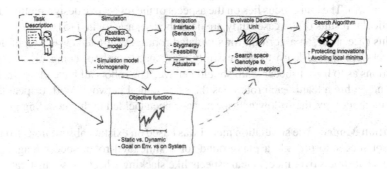

**Fig. 2.** Proposed design methodology

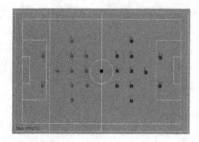 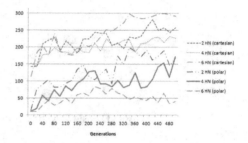

**Fig. 3.** Robot soccer game simulation starting situation erformance for different interface and decision unit configurations over the number of generations. Interface design has significant impact, while size of neural network does not. Plots refer to evolved fully meshed neural networks. Performance is measured by matching teams against each other in a full tournament.

The latter is focused on the actual representation of the components of the system. This might include one or more types of models depending on the homogeneity of the system. The reason why this part is separated from the system model is that the evolutionary method requires *evolvable* representations. There is, however, a vast set of such representations out there with their own advantages and disadvantages, so a careful decision has be made here.

In order for the evolutionary process to be operational there must be a valid and efficient search algorithm that iteratively optimizes the candidate solutions. While theoretically it can be completely separated from the representation of the evolvable components, it has to be noted that choosing the right algorithm has a significant overall effect on the convergence speed and on the quality of the results.

Last but not least the driving force of evolution must be also designed. While a good objective function has a key role in the process, there are unfortunately no guidelines available as to make a good function that not only ensures the desired behavior, but also makes the system avoiding unwanted states.

## 4   Case Study: Simulated Robot Soccer

The following case study describes an application of our design approach for a multi-robot scenario. The main focus lies on the aspects of the respective design methodology, the particular results of the case study are described separately in [15].

In this case study, an algorithm for simulated robot soccer teams was sought. Robot soccer players should learn to successfully play soccer without being given explicit instructions as to how a soccer team should behave. The robots all have the same hardware, and within a team, each robot has the same neural network. With respect to our design methodology, the following aspects have been tackled in the modeling process:

**Simulation Model:** The simulation model was based on existing abstractions from real robot soccer to provide a playground for a simulated robot soccer league. This model abstracts over mechanical aspects like slacking wheels or sophisticated sensor models for ball detection, while still providing a realistic interface for testing a distributed algorithm for making the robots play soccer. In particular, our simulation was based on the Robocup Soccer Simulation League without the features for centralized control via a coach.

**Interaction Interface:** Experiments with different set-ups showed that the sensor interface needs to be reduced to a few vectors indicating the position of the ball, the closest opponent and the closest teammate relative to the robot. In order to score a goal, the robots need to know where they are relative to the goal, so we added sensors indicating their relative field position. We tested the performance of evolved teams using a cartesian and a polar coordinate system representing the vectors. Results show that the cartesian coordinate system can be evolved to a better solution.

**Decision Unit:** Since there exists no particular theory, on how an evolvable decision unit for a soccer robot should be constructed, we experimented with several types of neural networks and varied the network size. Results indicate that a recursive neural network model, though being harder to analyze for a human observer, provides better capabilities than a feed forward model. The network size had less impact on the results than expected – solutions with 2, 4 or 6 neurons did not differ significantly in their performance (see Figure 3.

**Objective Function:** Playing soccer involves complex coordination of all team members based on variables of high dynamics like the position and momentum of the ball and all players. Therefore, we identified separate performance indicators, namely field coverage, ball possession, playing the ball actively, getting the ball near the opponents goal and scoring goals. Putting these factors hierarchically into an objective function enabled a smooth evolution of game-play [1]

A thorough description of the results is given in [15]. The robot soccer experiment was conducted with the tool FRamework for Evolutionary Design (FREVO) [16]. [2].

## 5   Conclusions and Future Work

This paper discusses the basic building blocks for applying evolutionary algorithms to the design of self-organizing systems for technical applications. Our survey of existing literature has revealed that despite the many reports on applying this approach to generate a self-organizing system, there is a need for a generic description of the underlying system engineering process. We therefore introduced several vital cornerstones of such an approach. In particular, we identified the task description, simulation setup, interaction interface, decision unit, search algorithms and fitness function.

The paper makes two main contributions: It supports the design process by defining the system aspects that must be considered. Its second contribution is that this approach may provide the basis for a qualified engineering process for self-organizing systems in technical applications. In future work we will elaborate our proposed methodology with detailed guidelines and principles for task description, simulation setup, interaction interface, decision unit, search algorithms and fitness function.

---

[1] See http://www.youtube.com/watch?v=cP035M_w82s for a video demo.

[2] FREVO including the robot soccer simulation described above are available as open source at http://frevo.sourceforge.net/

**Acknowledgments.** This work was supported by the Austrian Science Fund (FFG) grant FFG 2305537 and the Lakeside Labs GmbH, Klagenfurt, Austria, and funding from the European Regional Development Fund and the Carinthian Economic Promotion Fund (KWF) under grant KWF 20214|21532|32604, and KWF 20214|22935|34445. We would like to thank Lizzie Dawes for proofreading the paper.

# References

1. Elmenreich, W., de Meer, H.: Self-organizing networked systems for technical applications: A discussion on open issues. In: Hummel, K.A., Sterbenz, J.P.G. (eds.) IWSOS 2008. LNCS, vol. 5343, pp. 1–9. Springer, Heidelberg (2008)
2. Gershenson, C.: Design and Control of Self-organizing Systems. PhD thesis, Vrije Universiteit Brussel (2007)
3. Auer, C., Wüchner, P., de Meer, H.: A method to derive local interaction strategies for improving cooperation in self-organizing systems. In: Proceedings of the Third International Workshop on Self-Organizing Systems, Vienna, Austria (December 2008)
4. Quinn, M., Smith, L., Mayley, G., Husb, P.: Evolving teamwork and role allocation with real robots. In: Proceedings of the 8th International Conference on Artificial Life, pp. 302–311. MIT Press (2002)
5. Arteconi, S.: Evolutionary methods for self-organizing cooperation in peer-to-peer networks. Technical Report UBLCS-2008-5, Department of Computer Science, University of Bologna (2008)
6. Donaldson, M.C., Lachmann, M., Bergstrom, C.T.: The evolution of functionally referential meaning in a structured world. Journal of Theoretical Biology 246, 225–233 (2007)
7. Cucker, F., Smale, S.: The mathematics of emergence. The Japanese Journal of Mathematics 2, 197–227 (2007)
8. Licata, I., Sakaji, A. (eds.): Physics of Emergence and Organization. World Scientific (2008)
9. Fehr, E., Gintis, H.: Human nature and social cooperation. Annual Review of Sociology 33(3), 1–22 (2007)
10. Axtell, R.: The emergence of firms in a population of agents: Local increasing returns, unstable nash equilibria, and power law size distributions. Technical Report CSED Working Paper #3, Brookings Institution (June 1999)
11. Van Belle, T., Ackley, D.H.: Code factoring and the evolution of evolvability. In: Proceedings of the Genetic and Evolutionary Computation Conference, GECCO (2002)
12. Shoham, Y., Leyton-Brown, K.: Multiagent Systems: Algorithmic, Game-Theoretic, and Logical Foundations. Cambridge University Press (2008)
13. Koza, J.R.: Genetic programming: on the programming of computers by means of natural selection. MIT Press, Cambridge (1992)
14. Renz, W., Preisler, T., Sudeikat, J.: Mesoscopic stochastic models for validating self-organizing multi-agent systems. In: Proceedings of the 1st International Workshop on Evaluation for Self-Adaptive and Self-Organizing Systems, Lyon, France (September 2012)
15. Fehervari, I., Elmenreich, W.: Evolving neural network controllers for a team of self-organizing robots. Journal of Robotics (2010)
16. Sobe, A., Fehérvári, I., Elmenreich, W.: FREVO: A tool for evolving and evaluating self-organizing systems. In: Proceedings of the 1st International Workshop on Evaluation for Self-Adaptive and Self-Organizing Systems, Lyon, France (September 2012)

# Self-organization Promotes the Evolution of Cooperation with Cultural Propagation

Luis Enrique Cortés-Berrueco[1], Carlos Gershenson[2,4], and Christopher R. Stephens[3,4]

[1] Graduate Program in Computer Science and Engineering, Universidad Nacional Autónoma de México, Ciudad Universitaria , A.P.70-600, México D.F., México
lecortesb@comunidad.unam.mx

[2] Computer Science Department, Instituto de Investigaciones en Matemáticas Aplicadas y en Sistemas, Universidad Nacional Autónoma de México, Ciudad Universitaria, A.P.20-726, México D.F., México
cgg@unam.mx

[3] Gravitation Department. Instituto de Ciencias Nucleares, Universidad Nacional Autónoma de México, Ciudad Universitaria, A.P.70-543, México D.F., México
stephens@nucleares.unam.mx

[4] Centro de Ciencias de la Complejidad,
Universidad Nacional Autónoma de México

**Abstract.** In this paper three computational models for the study of the evolution of cooperation under cultural propagation are studied: Kin Selection, Direct Reciprocity and Indirect Reciprocity. Two analyzes are reported, one comparing their behavior between them and a second one identifying the impact that different parameters have in the model dynamics. The results of these analyzes illustrate how game transitions may occur depending of some parameters within the models and also explain how agents adapt to these transitions by individually choosing their attachment to a cooperative attitude. These parameters regulate how cooperation can self-organize under different circumstances. The emergence of the evolution of cooperation as a result of the agent's adapting processes is also discussed.

## 1    Introduction

Urban traffic problems have a complex behavior even in their most simplistic abstractions [2]. This behavior becomes more complex when driver interaction is added [10]. With these two premises, in this paper, three computational models, which were originally intended for studying driving behaviors and their impact in traffic, are presented. The aim is to explore the possibilities for controlling some of the complex characteristics of traffic, e.g. the consequences of driver interactions. The possibility of using them for other social problems is also discussed.

The models were conceived to study the evolution of cooperation [1], which studies how cooperative behaviors are preserved when selfish behaviors offer greater individual rewards, e.g. drivers who yield at lane changes or crossings, animal females taking care of their offspring, students who don't betray each other. Several

W. Elmenreich, F. Dressler, and V. Loreto (Eds.): IWSOS 2013, LNCS 8221, pp. 145–150, 2014.

abstractions for simulating different circumstances in which this phenomenon occurs have been proposed [6]. The models are based in three abstractions of different observed situations in which cooperation evolves [4]. Table 1 shows these abstractions as theoretical games in which players may choose between two behaviors: cooperative or selfish. When a player chooses to cooperate she has to pay a cost that her game partner is going to receive as a benefit. In each game, the decision of the players is influenced by a different probabilistic variable. This variable represents some feature that is exploited by the global behavior to favor the evolution of cooperation. For the kin selection strategy, relationships are exploited, the relation may be genetic, emotional or of any other type but as long as it its closer, cooperation will be more frequent. In the direct reciprocity case, the evolution of cooperation is linked with the probability of one player to play again with the same partner, and this condition is set because players are responding cooperation with cooperation and defection with defection. The feature exploited by the indirect reciprocity case is peer pressure. While more players get to know the actions of a determined player, she will obtain more benefits by cooperating and while fewer players get to know her actions more benefits she will obtain by defecting. Each strategy has a condition for the cooperation to become an evolutionary stable strategy (ESS, cooperators survive), another when cooperation becomes risk-dominant (RD, cooperators are near 1/2 of the population) and other when cooperators are advantageous (AD, cooperators are near 2/3 of the population).

**Table 1.** Rules for the three game strategies [4]

| Strategy | Payoff table | | | | ESS | RD | AD | Variables |
|---|---|---|---|---|---|---|---|---|
| Kin Selection | | | C | D | $\dfrac{b}{c} > \dfrac{1}{r}$ | $\dfrac{b}{c} > \dfrac{1}{r}$ | $\dfrac{b}{c} > \dfrac{1}{r}$ | b=benefit c=cost r=relatedness probability |
| | | C | (b-c)(1+r) | br-c | | | | |
| | | D | b-rc | 0 | | | | |
| Direct Reciprocity | | | C | D | $\dfrac{b}{c} > \dfrac{1}{w}$ | $\dfrac{b}{c} > \dfrac{2-w}{w}$ | $\dfrac{b}{c} > \dfrac{3-2w}{w}$ | b=benefit c=cost w=probability of next round |
| | | C | (b-c)/(1-w) | -c | | | | |
| | | D | b | 0 | | | | |
| Indirect Reciprocity | | | C | D | $\dfrac{b}{c} > \dfrac{1}{q}$ | $\dfrac{b}{c} > \dfrac{2-q}{q}$ | $\dfrac{b}{c} > \dfrac{3-2q}{q}$ | b=benefit c=cost q=social acquaintanceship |
| | | C | b-c | -c(1-q) | | | | |
| | | D | c(1-q) | 0 | | | | |

## 2     The Models

We developed agent-based computational models, using the NetLogo [12] platform, to better understand three strategies for the evolution of cooperation: kin selection, direct reciprocity and indirect reciprocity [4]. Unlike the results shown in [4] these models are focused in cultural evolution (players may choose to cooperate or not under different circumstances instead of been born as players who always cooperate or always defect), therefore, the first characteristic these models share is that they have a constant population. The behavior of the agents is similar to the one of agents described in Skyrm's Matching Pennies model [8]. Agents have a cooperation probability variable (*cp*) that determines their attachment to a cooperator strategy, while this variable gets a greater value the agent will cooperate more frequently. We will use this variable for deciding if the agents will cooperate or not in a particular game. Each agent is able to identify herself as a cooperator or as a defector:

$$\text{If } cp(p_i) > 0.5 \text{ then } c_i = p_i \text{ and } d_i = 0$$

$$\text{If } cp(p_i) \leq 0.5 \text{ then } d_i = p_i \text{ and } c_i = 0$$

Where $p_i$ is the i[th] agent or player, $c_i$ is the i[th] cooperator and $d_i$ is the i[th] defector.

As a consequence of this characteristic, a partition of the population can be made. The *initial proportions of cooperative versus defective* agents in the population (*ipc* and *ipd*) are relevant parameters of the model although they can be reduced to one as *ipc*=1-*ipd*. In order to identify agents as cooperators or defectors at the beginning of the simulations we give them an initial *cp*, thus, *initial cp of defectors (iicpd)* and *initial cp of cooperators (icpc)* are two more parameters. The agents move through a defined two-dimensional space. It has been seen [3] that in such cases population density is a key determinant of the dynamics, so the number of players (*population*) interacting in the defined space is another important parameter of the model.

Agents interact following the payoff tables corresponding to each strategy studied. As can be seen in Table 1, the three strategies share two variables: the benefit (*b*) one agent gives to another while cooperating and the cost (*c*) paid by the cooperator while cooperating. With the purpose of keeping track of the player's decisions, each one of them is assigned a *fitness* value from where the cost will be subtracted and the benefits will be added.

Appendix A of [13] shows the instructions executed by the agents during each iteration of the simulations. The behaviors of each game are detailed in appendix B of [13].

## 2.1 Tune Cooperation Probability

Agents adapt to the environment by modifying whether they are going to cooperate or defect in the next round. We implemented several exogenous tuning algorithms based on the literature [8][9][6][7][11] with no satisfactory results. We developed two self-organizing tuning alternatives and were tested with very good results:

- **Selfish fitness**: the agents only take their own *fitness* value as a parameter, unlike other algorithms that take into account the values of all the other players or values of the game partners. If the player cooperates or defects, and her *fitness* increases, then the agent increases or decreases her *cp* respectively. If the player cooperates or defects, and her *fitness* is reduced, then the agent decreases or increases her *cp* respectively. And if there is no change in the agent's fitness, then there is no change in her cp.

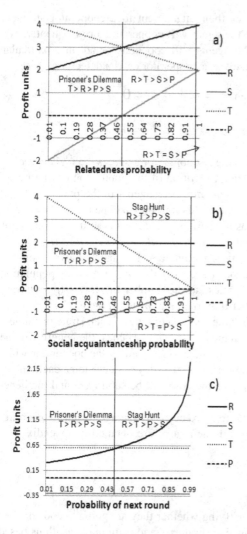

- **Selfish profit**: this is very similar to the previous one, but instead of comparing fitness values (sum of all profits obtained), agents act based in the comparison of the profit of their last game and the profit of the current round. Thus, if the player cooperates or defects and the actual profit is greater than the last one, then the agent increases or decreases her $cp$. If the player cooperates or defects and the actual profit is lower than the last one, then the agent decreases or increases her $cp$. Finally if the last profit is equal to the actual profit, then there is no change in the $cp$ value.

## 3    Methods

Before general results were obtained, an analysis of the variables was conducted. A primary goal of this analysis was to determine the impact of the parameters on the dynamics of the models. A secondary goal was to obtain parameter values that best exemplify the model's behavior. Detailed information about the experiments may be found in appendix C of [13].

**Fig. 1.** Graphics of the payoff table behaviors for x's values in the range [0,1] for a) Kin Selection, b) Indirect Reciprocity and c) Direct Reciprocity (DR is in logarithmic scale)

## 4    Results

Interesting results can be derived from a careful analysis of the graphics of the payoff table behavior as the corresponding probability takes values within the range [0,1] (Figure 1). As $x$ has a higher value, the agents move to different games. Using the same notations as in [6] that designates R (reward) when two agents cooperate, P (punishment) when two players defect, T (temptation) when one player defects and the other cooperates and S (sucker's payoff) when one player cooperates and the other defects; the move from game to game is described in appendix D of [13]. The values of $b$ and $c$ also determine the transition value of $x$ that takes the agents from one game to another. It is important to notice that the agents respond to these transitions by interacting with the others and adapt by using only simple local rules with only two basic requirements: 1) the agents must know how to distinguish that an amount is bigger than other and 2) the agents must have the notion of more is better. Within the strategies studied, other requirements are implicitly given for each case and are described in the appendix E of [13].

## 4.1    ESS, RD and AD Conditions

It was shown in Table 1, when the $x$ variable reaches certain value, ESS, RD and AD behaviors emerge. Figure 2 shows the characteristic behavior obtained by the analysis described in the behavior part of the Methods section; in it, these expected behaviors are not noticeable. Analyzing the model variables, we could find that these conditions are preserved, but the effect can be observed by setting the starting population of cooperators as 2/3, 1/2 and 1/3 of the total population for ESS, RD and AD respectively and obtaining a graphic similar to Figure 2. This result is shown in appendix F of [13] along with the impact of all the parameters to each strategy.

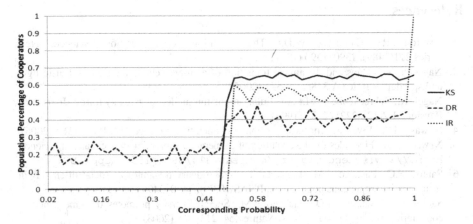

**Fig. 2.** Percentage of cooperating population while the corresponding probability is varied

## 4.2    Self-organization

Self-organization seems to be a key element for the cultural evolution of cooperation. There are several studies [8][9][6][7][11] in which the agents just can choose between cooperate or defect. In the models presented here, the players can choose the degree of cooperation that they consider the best. Another important difference between our models and the others [6][7] is the inclusion of a probabilistic variable so that agents choose the degree of cooperation by comparing their actual state with a past one instead of comparing their state with the state of others. This is important because each agent may be seen as an individual system adaptable to different environments or contexts using the information given by the interactions with other similar, not necessarily equal, systems.

## 5 Conclusions

Three agent-based computational models for the study of the evolution of cooperation under cultural propagation were described. It was shown that their behavior is the result of transitions between games defined by Game Theory. The transitions are consequences of the structure determined by the payoff matrixes of the three strategies studied. Each of these strategies abstracts well-known real behaviors [4]; hence the importance of creating computational models that let us experiment exhaustively different circumstances for these phenomena, a difficult task with living organisms. The impact of the parameters in each model was analyzed to better understand how to manipulate the models and adapt them for more specific studies, such as social phenomena (e.g. bulling, market formation, and migration), traffic problems (e.g. signaling issues, conflicting use of roads) and biological processes (e.g. interactions between organisms and populations).

## References

1. Axelrod, R., Hamilton, W.D.: The evolution of cooperation. Science, New Series 211(4489), 1390–1396 (1981)
2. Nagatani, T.: Chaos and dynamical transition of a single vehicle induced by traffic light and speedup. Physica A 348, 561–571 (2005)
3. Nagel, K., Schreckenberg, M.: A cellular automaton model for freeway traffic. Journal de Physique 1 2(12), 2221–2229 (1992)
4. Nowak, M.A.: Five rules for the evolution of cooperation. Science 314, 1560–1563 (2006)
5. Nowak, M.A.: Five rules for the evolution of cooperation supporting online material. Science, http://www.sciencemag.org/content/314/5805/1560/suppl/DC1
6. Santos, F.C., Pacheco, J.M., Lenaerts, T.: Evolutionary dynamics of social dilemmas in structured heterogeneous populations. PNAS 103(9), 3490–3494 (2006)
7. Santos, F.C., Santos, M.D., Pacheco, J.M.: Social diversity promotes the emergence of cooperation in public goods games. Nature 454, 213–216 (2008)
8. Skyrm, B.: The dynamics of rational deliberation. Hardvard University Press, Cambridge (1990)
9. Skyrms, B.: Dynamic models of deliberation and the theory of games. In: Proceeding of the 3rd Conference of Theoretical Aspects of Reasoning about Knowledge, pp. 185–200 (1990)
10. Wastavino, L.A., Toledo, B.A., Rogan, J., Zarama, R., Muñoz, V., Valdivia, J.A.: Modeling traffic on crossroads. Physica A 381, 411–419 (2007)
11. Weirich, P.: Computer simulations in game theory (2006), http://philsci-archive.pitt.edu/2754/
12. Wilensky, U.: NetLogo. Center for Connected Learning and Computer-Based Modeling, Northwestern University, Evanston, IL (1999), http://ccl.northwestern.edu/netlogo/
13. arXiv:1302.6533 [cs.GT]

# Not All Paths Lead to Rome:
# Analysing the Network of Sister Cities

Andreas Kaltenbrunner*, Pablo Aragón, David Laniado, and Yana Volkovich**

Barcelona Media Foundation, Barcelona, Spain
{name.surname}@barcelonamedia.org

**Abstract.** This work analyses the practice of sister city pairing. We investigate structural properties of the resulting city and country networks and present rankings of the most central nodes in these networks. We identify different country clusters and find that the practice of sister city pairing is not influenced by geographical proximity but results in highly assortative networks.

**Keywords:** Social network analysis, sister cities, social self-organisation.

## 1 Introduction

Human social activity in the form of person-to-person interactions has been studied and analysed in many contexts, both for online [7] and off-line behaviour [11]. However, sometimes social interactions give rise to relations not anymore between individuals but rather between entities like companies [4], associations [8] or countries [2]. Often these relations are associated with economic exchanges [2], sports rivalry [9] or even cooperation [8].

In this work we study one type of such relations expressed in the form of *sister city partnerships*[1]. The concept of sister cities refers to a partnership between two cities or towns with the aim of cultural and economical exchange. Most partnerships connect cities in different countries, however also intra-country city partnerships exist. Our study aims at understanding some of the basic social, geographical and economic mechanisms behind the practice of city pairings.

We extracted the network of sister cites as reported on the English Wikipedia, as far as we know the most extensive but probably not complete collection of this kind of relationships. The resulting social network, an example of social self organisation, is analysed in its original form and aggregated per country. Although there exist studies that analyse networks of cities (e.g. networks generated via aggregating individual phone call interactions [6]) to the best of our knowledge this is the first time that institutional relations between cities are investigated.

* This work was supported by Yahoo! Inc. as part of the CENIT program, project CEN-20101037, and ACC1Ó -Generalitat de Catalunya (TECRD12-1-0003).
** Acknowledges the Torres Quevedo Program from the Spanish Ministry of Science and Innovation, co-funded by the European Social Fund.
[1] Sometimes the same concept is also referred to as *twin town*, *partnership town*, *partner town* or *friendship town*. Here we use preferentially the term *sister city*.

W. Elmenreich, F. Dressler, and V. Loreto (Eds.): IWSOS 2013, LNCS 8221, pp. 151–156, 2014.

**Table 1.** Network properties: number of nodes $N$ and edges $K$, average clustering coefficient $\langle C \rangle$, % of nodes in the giant component GC, average path-length $\langle d \rangle$

| network | $N$ | $K$ | $\langle C \rangle$ | % GC | $\langle d \rangle$ |
|---|---|---|---|---|---|
| city network | 11 618 | 15 225 | 0.11 | 61.35% | 6.74 |
| country network | 207 | 2933 | 0.43 | 100% | 2.12 |

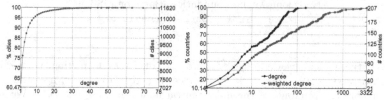

**Fig. 1.** Cumulative degree distribution in the city (left) and country networks (right)

## 2   Dataset Description

The dataset used in this study was constructed (using an automated parser and a manual cleaning process) from the listings of sister cities on the English Wikipedia.[2] We found 15 225 pairs of sister cities, which form an undirected[3] *city network* of 11 618 nodes. Using the `Google Maps API` we were able to geo-locate 11 483 of these cities.

We furthermore construct an aggregated undirected and weighted *country network*, where two countries $A$ and $B$ are connected if a city of country $A$ is twinned with a city of country $B$. The number of these international connections is the edge weight. The country network consists of 207 countries and 2 933 links. Some countries have self-connections (i.e. city partnerships within the same country). Germany has the largest number of such self links as a result of many sister city relations between the formerly separated East and West Germany.

Table 1 lists the principal macroscopic measures of these two networks. The clustering coefficient of the city network is comparable to the values observed in typical social networks [10]. Also the average path-length between two cities is with 6.7 in line with the famous six-degrees-of-separation. The country network is denser, witnessed by the remarkably high value of the clustering coefficient ($\langle C \rangle = 0.43$), and a very short average distance of 2.12.

In Figure 1 we plot the degree distributions of both networks. We observe in Figure 1 (left) that more than 60% of the cities have only one sister city, about 16% have two and only less than 4% have more than 10. For the countries we observe in Figure 1 (right) that around 58% of the countries have less than 10 links to other countries, but at the same time more than 20% of the countries have more than 100 sister city connections (i.e. weighted degree $\geq 100$). Both networks have skewed degree-distributions with a relative small number of hubs.

---

[2] Starting from `http://en.wikipedia.org/wiki/List_of_twin_towns_and_sister_cities`, which includes links to listings of sister cities grouped by continent, country and/or state.

[3] Although only 29.8% of the links were actually reported for both directions.

**Table 2.** Comparing assortativity coefficients $r$ of the city network with the mean assortativity coefficients $r_{rand}$ and the corresponding stdv $\sigma_{rand}$ of randomised networks. Resulting Z-scores $\geq 2$ (in bold) indicate assortative mixing. Apart from the city degrees, the city properties used coincide with the corresponding country indexes.

| property | $r$ | $r_{rand}$ | $\sigma_{rand}$ | $Z$ |
|---|---|---|---|---|
| city degree | 0.3407 | -0.0037 | 0.0076 | **45.52** |
| Gross Domestic Product (GDP)[4] | 0.0126 | -0.0005 | 0.0087 | 1.51 |
| GDP per capita[5] | 0.0777 | 0.0005 | 0.0078 | **9.86** |
| Human Development Index (HDI)[6] | 0.0630 | -0.0004 | 0.0075 | **8.46** |
| Political Stability Index[7] | 0.0626 | 0.0004 | 0.0090 | **6.94** |

## 3  Assortativity

To understand mixing preferences between cities, we follow the methodology of [3] and calculate an assortativity measure based on the Z-score of a comparison between the original sister city network and 100 randomised equivalents. For degree-assortativity, randomised networks are constructed by reshuffling the connections and preserving the degree; in the other cases, the network structure is preserved while the values of node properties are reshuffled.

Table 2 gives an overview of the results. We find that the city network is highly assortative indicating a clear preference for connections between cities with similar degree. We also analyse assortativity scores for other variables and find that cities from countries with similar Gross Domestic Product (GDP) per capita, Human Development Index or even similar indexes of political stability are more likely to twin. Only for the nominal GDP neutral mixing is observed.

## 4  Rankings

We discuss now city and country rankings based on centrality measures. For the sister city network we show the top 20 cities ranked by degree (Table 3, left). Saint Petersburg, often referred to as the geographic and cultural border of the West and East, is the most connected and also most central sister city. There are also cities, such as Buenos Aires, Beijing, Rio de Janeiro and Madrid, which have large degrees but exhibit lower betweenness ranks. In particular, the Spanish and the Chinese capitals have significantly lower values of betweenness, which could be caused by the fact that other important cities in these countries (e.g. Barcelona or Shanghai) act as primary international connectors.

In Table 3 (right) we present rankings for the country network. In this case the USA lead the rankings in the two centrality measures we report. The top ranks are nearly exclusively occupied by Group of Eight (G8) countries suggesting a relation between economic power and sister city connections.

---

[4] Source http://en.wikipedia.org/wiki/List_of_countries_by_GDP_(nominal)

[5] Source: http://en.wikipedia.org/wiki/List_of_countries_by_GDP_(nominal)_per_capita

[6] Source: http://en.wikipedia.org/wiki/List_of_countries_by_Human_Development_Index

[7] Source: http://viewswire.eiu.com/site_info.asp?info_name=social_unrest_table

**Table 3.** The top 20 cities (left) and countries (right) ranked by (weighted) degree. Ranks for betweenness centrality in parenthesis.

| city | degree | betweenness | | country | weighted degree | betweenness | |
|---|---|---|---|---|---|---|---|
| Saint Petersburg | 78 | 1 562 697.97 | (1) | USA | 4520 | 9855.74 | (1) |
| Shanghai | 75 | 825 512.69 | (4) | France | 3313 | 1946.26 | (3) |
| Istanbul | 69 | 601 099.50 | (12) | Germany | 2778 | 886.78 | (6) |
| Kiev | 63 | 758 725.12 | (5) | UK | 2318 | 2268.32 | (2) |
| Caracas | 59 | 430 330.45 | (23) | Russia | 1487 | 483.65 | (9) |
| Buenos Aires | 58 | 348 594.25 | (36) | Poland | 1144 | 34.09 | (33) |
| Beijing | 57 | 184 090.42 | (124) | Japan | 1131 | 168.47 | (20) |
| São Paulo | 55 | 427 457.92 | (24) | Italy | 1126 | 849.20 | (7) |
| Suzhou | 54 | 740 377.17 | (6) | China | 1076 | 1538.42 | (4) |
| Taipei | 53 | 486 042.21 | (20) | Ukraine | 946 | 89.22 | (27) |
| Izmir | 52 | 885 338.70 | (3) | Sweden | 684 | 324.84 | (14) |
| Bethlehem | 50 | 1 009 707.96 | (2) | Norway | 608 | 147.06 | (22) |
| Moscow | 49 | 553 678.88 | (16) | Spain | 587 | 429.79 | (11) |
| Odessa | 46 | 724 833.39 | (8) | Finland | 584 | 30.24 | (35) |
| Malchow | 46 | 519 872.56 | (17) | Brazil | 523 | 332.26 | (13) |
| Guadalajara | 44 | 678 060.06 | (9) | Mexico | 492 | 149.70 | (21) |
| Vilnius | 44 | 589 031.92 | (14) | Canada | 476 | 72.01 | (28) |
| Rio de Janeiro | 44 | 381 637.67 | (29) | Romania | 472 | 34.44 | (32) |
| Madrid | 40 | 135 935.80 | (203) | Belgium | 464 | 145.18 | (23) |
| Barcelona | 39 | 266 957.88 | (60) | The Netherlands | 461 | 274.79 | (16) |

## 5    Clustering of the Country Network

In Figure 2 we depict the country network. Node size corresponds to the weighted degree, and the width of a connection to the number of city partnerships between the connected countries. The figure shows the central position of countries like the USA, France, UK and China in this network.

The colours of the nodes correspond to the outcome of node clustering with the Louvain method. We find 4 clusters. The largest one (in violet) includes the USA, Spain and most South American, Asian, and African countries. The second largest (in green) is composed of Eastern-European and Balkan countries: Turkey, Russia, and Poland are the most linked among them. The third cluster (in red) consists of Central and Western-European countries and some of their former colonies. It is dominated by Germany, UK, France and the Netherlands. Finally, the smallest cluster (in cyan) mainly consists of Nordic countries.

The clustering suggests cultural or geographical proximity being a factor in city partnerships. In the next section we will investigate this further.

## 6    Distances

To test the extent to which geographical proximity is an important factor for city partnership we analyse the distributions of geographical distances between all pairs of connected sister cities.

Figure 3 depicts this distribution as a histogram (blue bars in the left sub-figure) or as a cumulative distribution (blue curve in the right sub-figure). The figure also shows the overall distance distribution between all possible pairs (connected or not) of geo-located sister cities (green bars and red curve). There is nearly no difference (apart from some random fluctuations) between these two

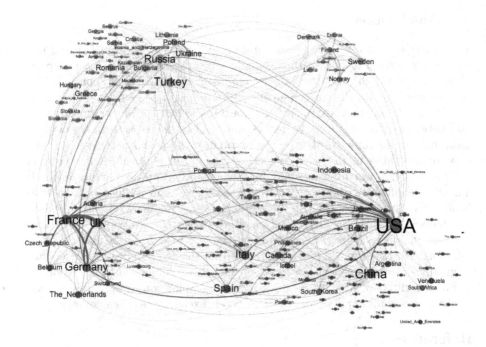

**Fig. 2.** Country network: node size corresponds to degree and node colours indicate the four clusters obtained with the Louvain method

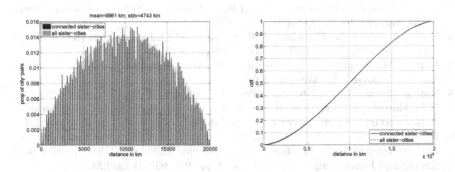

**Fig. 3.** Distribution of the distances between connected sister cities (blue) and the practically identical distance distribution between all cities (green in pdf, red in cdf)

distributions. The fluctuations vanish in the cumulative distributions where the two curves are nearly overlapping. Only for very close cities it is slightly more likely than expected by random choice to establish a city sistership. This can also be observed in the very small difference of the average distance of two randomly chosen cities (10 006 km) and a pair of connected sister cities (9 981 km).

# 7    Conclusions

We have studied the practice of establishing sister city connections from a network analysis point of view. Although there is no guarantee that our study covers all existing sister city relations, we are confident that the results obtained give reliable insights into the emerging network structures and country preferences.

We have found that sister city relationships reflect certain predilections in and between different cultural clusters, and lead to degree-assortative network structures comparable to other types of small-world social networks. We also observe assortative mixing with respect to economic or political country indexes.

The most noteworthy result may be that the geographical distance has only a negligible influence when a city selects a sister city. This is different from what is observed for person-to-person social relationships (see for example [5]) where the probability of a social connection decays with the geographical distance between the peers. It may, thus, represent the first evidence in real-world social relationships (albeit in its institutional form) for the death of distance, predicted originally as a consequence of decrease of the price of human communication [1].

Possible directions for future work include combination of the analysed networks with the networks of air traffic or goods exchange between countries.

# References

1. Cairncross, F.: The death of distance: How the communications revolution is changing our lives. Harvard Business Press (2001)
2. Caldarelli, G., Cristelli, M., Gabrielli, A., Pietronero, L., Scala, A., Tacchella, A.: A network analysis of countries export flows: Firm grounds for the building blocks of the economy. PLoS One 7(10), e47278 (2012)
3. Foster, J., Foster, D., Grassberger, P., Paczuski, M.: Edge direction and the structure of networks. PNAS 107(24), 10815–10820 (2010)
4. Höpner, M., Krempel, L.: The politics of the german company network. Competition and Change 8(4), 339–356 (2004)
5. Kaltenbrunner, A., Scellato, S., Volkovich, Y., Laniado, D., Currie, D., Jutemar, E.J., Mascolo, C.: Far from the eyes, close on the Web: impact of geographic distance on online social interactions. In: Proceedings of WOSN 2012. ACM (2012)
6. Krings, G., Calabrese, F., Ratti, C., Blondel, V.: Scaling behaviors in the communication network between cities. In: International Conference on Computational Science and Engineering, CSE 2009, vol. 4, pp. 936–939. IEEE (2009)
7. Mislove, A., Marcon, M., Gummadi, K.P., Druschel, P., Bhattacharjee, B.: Measurement and analysis of online social networks. In: Proc. of IMC (2007)
8. Moore, S., Eng, E., Daniel, M.: International NGOs and the role of network centrality in humanitarian aid operations: A case study of coordination during the 2000 Mozambique floods. Disasters 27(4), 305–318 (2003)
9. Mukherjee, S.: Identifying the greatest team and captaina complex network approach to cricket matches. Physica A (2012)
10. Newman, M., Watts, D., Strogatz, S.: Random graph models of social networks. PNAS 99(suppl. 1), 2566–2572 (2002)
11. Wasserman, S., Faust, K.: Social network analysis: Methods and applications, vol. 8. Cambridge University Press (1994)

# A Hybrid Threat Detection and Security Adaptation System for Industrial Wireless Sensor Networks

Mohammed Bahria, Alexis Olivereau, and Aymen Boudguiga

CEA, LIST, Communicating Systems Laboratory, Gif-sur-Yvette - France
{mohammed.bahria,alexis.olivereau,aymen.boudguiga}@cea.fr

**Abstract.** Wireless Sensor Networks (WSNs) led the way to new forms of communications, which extend today the Internet paradigm to unforeseen boundaries. The legacy industry, however, is slower to adopt this technology, mainly for security reasons. Self-managed security systems allowing a quicker detection of and better resilience to attacks, may counterbalance this reluctance. We propose in this paper a hybrid threat detection and security adaptation system, designed to run on top of industrial WSNs. We explain why this system is suitable for architectures mainly composed of constrained or sleeping devices, while being able to achieve a fair level of autonomous security.

**Keywords:** Threat detection, sensor network, security adaptation.

## 1 Introduction

The gain of maturity of WSN technologies accelerates their adoption in the industry, and this adoption is all the quicker as WSNs answer to classical needs of industrial scenarios: physical values monitoring, asset supervision and facilities surveillance are all key requirements in these scenarios, for which dedicated sensors are available. However, even though cost effective devices and energy-efficient technologies and protocols are available, the underlying security question impedes the use of WSNs in the most critical industrial scenarios. The inherent vulnerability of WSN nodes, due to their exposed location and their use of wireless communications, is such that a WSN has to mimic all security features from the legacy Internet, while also adding specific use cases and taking into account the strong shortcomings of the WSN nodes.

In this paper, we introduce a new Threat Detection (TD) system that is lightweight enough to be run on sensor nodes. We couple it with a flexible Security Adaptation (SA) system, which dynamically updates the security policies in special places in the network. We show that the use of this hybrid threat detection/reaction system greatly improves the resilience of the industrial WSN, without bringing in excessive energy consumption. The proposed solution is based on a partly centralized architecture and specifies new roles for WSN entities, in accordance with their status and capabilities.

W. Elmenreich, F. Dressler, and V. Loreto (Eds.): IWSOS 2013, LNCS 8221, pp. 157–162, 2014.

## 2    Problem Statement

Both threat detection and security adaptation raise issues with respect to their adaptability to WSNs. Threat detection challenges the constrained nodes' limited energy resources: it involves both costly [3] passive monitoring, and heavy signature-based threat identification. Security adaptation challenges the inherent heterogeneity of a wireless sensor network: certain constrained nodes may be unable to comply with a new security policy. The sleeping/awake cycle of sensor nodes makes the operation of both security subsystems more complex, introducing desynchronization in it.

A specific factor related to the industrial scenario is the WSN real-world topology. An industrial facility is made of multiple distinct zones, such as the external perimeter, the administrative building, the workshop and the storage area. All of these zones feature different sensors and are characterized by different criticality levels. As an example, Figure 1 represents a schematized industrial network made of a production facility and an office building, both being protected by a fence. Each zone is equipped with sensors of two kinds, relevant to the concerned area.

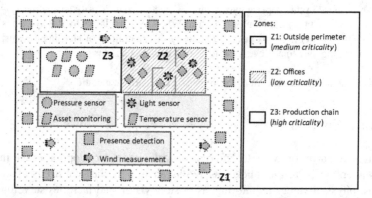

**Fig. 1.** Example of an industrial wireless sensor network

## 3    Related Work

Roman et al. [4], define a threat detection system based on monitoring agents. They distinguish between local agents (able to process only the packet they actively forward or to act as spontaneous watchdogs with a one-in-$n$ probability, $n$ being the number of neighbors) and global agents. Loannis et al. [5] also apply the watchdog behavior, with a higher emphasis on countering malicious monitoring nodes. Huang and Lee [6] propose an energy-efficient monitoring system. They select a single node, designated through a fair election process, to perform monitoring at a given time within a given cluster. Though, the election process is heavy for constrained nodes.

To summarize, [4] and [5] misuse the watchdog behavior while [6] uses a costly election process, leading to expensive message exchange.

Ma et al. [7] propose a Self-Adaptive Intrusion Detection System in WSN (SAID). Their approach does not take into the account the global view of the network, which may lead to incoherence. In addition, they propose to apply a new policy without checking if the nodes can handle it, which may be harmful to the network. Lacoste et al. [8] propose an approach that uses context-awareness for adaptive security. The context knowledge is combined with confidence and reputation metadata. The disadvantage of this approach is the expensive exchange of messages to maintain the coherence of the metadata. Younis et al. [9] suggest an Adaptive Security Provision (ASP), which adjusts security packet security processing based on trust and threats associated to routes. This solution is however too heavy for WSN. In addition, if a node that belongs to a route crashes, the route must be removed and new one will have to be computed. Finally, this solution is only suitable for routing security. Taddeo et al. [10] propose a method that permits the self-adaptation of security mechanisms. However, they always start with the highest security level, which could be costly for WSN nodes. Adaptive Security System (ADAPSEC) is reconfigurable security architecture for WSNs that has been developed by Shi et al. [11]. However it assumes both constant monitoring by each node and local inferring of new policies that the node should apply upon attack detection. Both would lead to high resource consumption. M.-Y. Hsieh et al. [12] base the WSN security on a trust management system. This approach delays the threat identification, though.

In addition to the identified shortcomings, neither of the previous security adaptation systems takes into consideration the sleep mode, although it is of paramount importance in WSN: a sensor node spends most of the time in sleep mode, wakes up to collect information and push it towards a sink or server, and reverts back to sleep.

## 4    Solution Description

### 4.1    Assumptions, Components and Roles

The network is supposed to be divided into zones, each containing one or more sensor clusters. Zones and clusters have different criticalities, security levels and security policies. It is also assumed that the sensors in one cluster can communicate with each other. In addition, we assume that the awake time is negligible compared to the sleep time. Finally, we exclude any form of synchronization between nodes.

The security system we present in this paper is made up of the following elements:

- *Threat Detection Client*, which identifies threats and notifies the TD server.
- *Threat Detection Server*, which chooses which sensor(s) will be in monitoring mode for each cluster by taking into account status parameters such as batteries level and available resources, updates the global network database, receives the alarms from TD clients and transfers them to the SA Server.
- *Security Adaptation Server*, which decides upon threat identification which is the best policy to apply and stores the new policy in the security policy mailbox.
- *Security Adaptation Client*, which regularly prompts the SA Server for new policies, and either applies them or generates new affordable security policies.

- *Inference engine*, which deals with reasoning and allows for easy rule changes.
- *Security Policy mailbox*, which stores the generated policies and delivers them at nodes request. The use of this module is required since the nodes are not synchronized. It also reduces the overall bandwidth consumption.
- *Global Network Database*, which contains a global view of the network and the threats detected in the past.

## 4.2    Operation

A node joining the network is in **Bootstrapping** mode. The newly joining node first sends a registration request to the TD server, informing this latter about its potential monitoring abilities. The TD server then registers the node and responds with a configuration message specifying whether it should remain in normal mode or temporarily switch to monitoring mode. The decision by the TD is based on its knowledge of current and, in some cases, foreseen contexts of the candidate monitoring nodes. This contextual information includes data related to the nodes resources (e.g. battery level), locations, and capabilities (e.g. number of observables neighbours). With this information, the TD server can identify the best node in the cluster for acting as a monitoring entity, and configure it with this role for a certain period of time. Once the monitoring delay expires, the TD server designates a new cluster monitoring node.

A node switches to **Monitoring Mode** when ordered to do so by the TD server. The sequence of actions performed when in monitoring node is:

1. When the TD Client detects a threat, it sends an alarm to the TD server that includes information about the threat. This information contains at least the IP addresses of the attacker and target(s) and the type of attack.
2. Upon receiving the alarm, the TD Server reports it to the SA server, optionally after having aggregated multiple alarm messages and/or having assessed the quality of the evaluator. The SA Server then uses the inference engine to determine which policies have to be applied to counter the detected threat. Next, it stores the new policies in the security policy mailbox. Depending on the global state of the network and type of the threat, a new policy may be wide-scale or local.
3. In monitoring mode, the TD client on the sensor regularly polls the security policy mailbox by sending a dedicated inquiry message to the SA server.
4. The SA server sends the requested new policy if it exists. Otherwise it replies with a message telling the node that it is not to enforce a new policy.
5. If a new policy is received, the SA Client checks whether it can be enforced by checking the available resources and safety constraints. If it finds out that applying the policy would be too costly or would put the safety or workers/facilities at risk, it tries to find a trade off in the form of a less demanding security policy.
6. The SA client configures the security services in accordance with the received or self-determined security policy. It then sends an ACK if the received policy was applied without change, or sends a descriptor of the locally generated policy.
7. The SA Server receives the ACK or locally generated policy descriptor and updates the global network database accordingly.

The **Normal Mode** is the default mode for a bootstrapped sensor that has not been designated as a monitoring node. In this mode, the sensors alternate between active and sleeping states. Upon leaving sleeping state, the node interrogates the SA server about an eventual new policy to enforce. It then performs the task(s) for which it has left the sleeping state. An alarm may be raised by a node in Normal Mode only if one of the run tasks detects a threat and notifies the TD Client through an API call.

The overall process of our solution, depicting its state machines and internal/network message exchanges, is depicted in Figure 6.

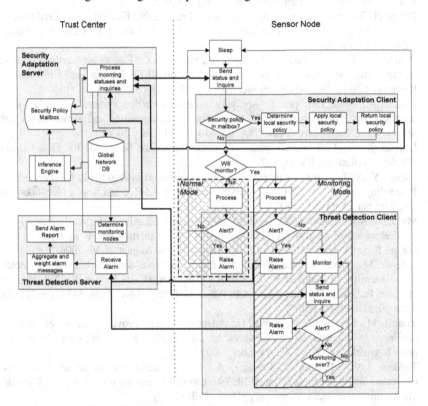

**Fig. 2.** Overall logical architecture and state machines

# 5    Conclusion

This paper presents an adaptive autonomous security system for industrial wireless sensor networks that features both threat detection and adaptive security. Both of these subsystems involve semi-centralized processes. The switch from normal threat detection mode to monitoring mode is triggered by the threat detection server, which bases on regular reports from nodes and updates its decision accordingly. The security adaptation system relies on server-issued policies, but the last word on how to enforce these policies remains with the sensor nodes. This system is currently being implemented for multiple industrial scenarios.

**Acknowledgement.** This work was financially supported by the EC under grant agreement FP7-ICT-258280 TWISNet project.

# References

1. Refaei, M.T., Srivastava, V., DaSilva, L., Eltoweissy, M.: A reputation-based mechanism for isolating selfish nodes in ad hoc networks. In: Mobile and Ubiquitous Systems: Networking and Services, MobiQuitous 2005 (July 2005)
2. Debar, H., Thomas, Y., Cuppens, F., Cuppens-Boulahia, N.: Enabling automated threat response through theuse of a dynamic security policy. Journal in Computer Virology (JCV) 3 (August 2007)
3. Wander, A., Gura, N., Eberle, H., Gupta, V., Shantz, S.C.: Energy analysis of public-key cryptography for wireless sensor networks. In: Third IEEE International Conference on Pervasive Computing and Communications, pp. 324–328 (2005)
4. Roman, R., Zhou, J., Lopez, J.: Applying intrusion detection systems to wireless sensor networks. In: Proceedings of IEEE Consumer Communications and Networking Conference (CCNC 2006), Las Vegas, USA, pp. 640–644 (January 2006)
5. Ioannis, K., et al.: Toward Intrusion Detection in Sensor Networks. In: 13th European Wireless Conference, Paris (2007)
6. Huang, Y.-A., Lee, W.: A cooperative intrusion detection system for ad hoc networks. In: Proceedings of the 1st ACM Workshop on Security of Ad Hoc and Sensor Networks, Fairfax, Virginia, October 31 (2003)
7. Ma, J., Zhang, S., Zhong, Y., Tong, X.: SAID: A self-adaptive intrusion detection system in wireless sensor networks. In: Proceedings of the 7th International Conference on Information Security Applications: Part I, Jeju Island, Korea, August 28-30 (2006)
8. Lacoste, M., Privat, G., Ramparany, F.: Evaluating confidence in context for context-aware security. In: Schiele, B., Dey, A.K., Gellersen, H., de Ruyter, B., Tscheligi, M., Wichert, R., Aarts, E., Buchmann, A.P. (eds.) AmI 2007. LNCS, vol. 4794, pp. 211–229. Springer, Heidelberg (2007)
9. Younis, M., Krajewski, N., Farrag, O.: Adaptive security provision for increased energy efficiency in Wireless Sensor Networks. In: 2009 IEEE 34th Conference on Local Computer Networks, pp. 999–1005 (October 2009)
10. Taddeo, A.V., Micconi, L., Ferrante, A.: Gradual adaptation of security for sensor networks. In: Proceedings of the IEEE International Symposium on a World of Wireless Mobile and Multimedia Networks, pp. 1–9. IEEE (2010)
11. Shi, K., Qin, X., Cheng, Q., Cheng, Y.: Designing a Reconfigurable Security Architecture for Wireless Sensor Networks. In: World Congress on Software Engineering, pp. 154–158. IEEE (2009)
12. Hsieh, M.-Y., Huang, Y.-M., Chao, H.-C.: Adaptive security design with malicious node detection in cluster-based sensor networks. Computer Communications 30(11-12), 2385–2400 (2007)

# Mapping of Self-organization Properties and Non-functional Requirements in Smart Grids[*]

Sebastian Lehnhoff[1], Sebastian Rohjans[1], Richard Holzer[2],
Florian Niedermeier[2], and Hermann de Meer[2]

[1] OFFIS – Institute for Information Technology, Escherweg 2, 26121 Oldenburg,
Germany
firstname.lastname@offis.de

[2] Faculty of Computer Science and Mathematics, University of Passau, Innstraße 43,
94032 Passau, Germany
firstname.lastname@uni-passau.de

**Abstract.** Future electrical power networks will be composed of large collections of autonomous components. Self-organization is an organizational concept that promises robust systems with the ability to adapt themselves to system perturbations and failures and thus may yield highly robust systems with the ability to scale freely to almost any size. In this position paper the authors describe the well-established process of use case based derivation of non-functional requirements in energy systems and propose a mapping strategy for aligning properties of self-organizing systems with the ICT- and automation system requirements. It is the strong belief of the authors that such a mapping will be a key factor in creating acceptance of and establishing self-organization in the domain of electrical energy systems.

**Keywords:** Smart Grid, Self-organization, Quality of Service, Quantitative Measures.

## 1 Introduction

In order to assess the requirements for ICT- and automation systems that have to support and enable a Smart Grid composed of large collections of autonomous components to perform appropriate functions a use case based requirements engineering methodology can be applied based on IEC PAS 62559 [1]. Here, use cases are established for the purpose of identifying how a system should behave in relation to its actors and components in order to ensure a specified task or function e.g. in due time, with the necessary precision, while meeting certain service level agreements etc. Thus, use cases may document non-functional requirements for usage scenarios of future electrical power networks and support technology decisions in order to implement extensive and complex Smart Grids.

---

[*] This research is partially supported by SOCIONICAL (FP7, ICT-2007-3-231288), by All4Green (FP7, ICT-2011-7-288674) and by the NoE EINS (FP7, ICT-2011-7-288021).

W. Elmenreich, F. Dressler, and V. Loreto (Eds.): IWSOS 2013, LNCS 8221, pp. 163–168, 2014.

In this position paper we propose a mapping strategy for aligning properties of self-organizing systems with the ICT- and automation system Quality of Service (QoS) requirements.

## 2 Use Case Based Requirements Engineering in Smart Grids

In order to derive what kind of functions to implement and what kind of systems to build for executing a future Smart Grid Use Case a number of questions need to be addressed [2]. *Functional* issues may address the system's flexibility for e.g. providing some kind of power reserve for load balancing, or the system's capability to detect certain phenomena that may be characteristic of a stability issue (if these are part of the given use case). However, for the development of a proper supporting ICT-system *non-functional* requirements may be more relevant and address issues like safety, security, performance or accuracy of certain measurements. In order to manage these kinds of issues and questions the methodology utilizes a reference architecture to identify participating system components and their interfaces to each other. A properly defined use case should then be able to identify relevant actors and components consistently from within this reference architecture and specify their interactions with each other. The European Union has issued a mandate (M/490) for the standardization of Smart Grid functionalities and use cases [3] to three European standards organizations (CEN, CENELEC and ETSI) to develop a common framework for representing functional information data flows between power system domains and their relevant subdomains as well as their architectures (the Smart Grid Architecture Model - SGAM) [4]. Additionally, a sustainable process and method is developed for use case specification and formalization allowing stakeholder interactions, gap analysis and (for the scope of this paper most important) the derivation of non-functional requirements necessary for the successful execution of a given task.

### 2.1 The Smart Grid Reference Architecture

The ISO/IEC 42010 defines a reference architecture as a description of a system's structure in terms of interactions between its element types and their environment [5]. The SGAM is the reference architecture for describing the Smart Grid and especially ICT- and automation systems within this domain. Starting from a contextual "component layer" spanning the power system domains in terms of the energy conversion chain and its equipment against the hierarchical levels (zones) of power system management (see bottom layer in Figure 1). The purpose of this layer is to emphasize the physical distribution of all participating components including actors, applications, power system equipment (at process and field level), protection and remote-control devices, communication infrastructure as well as any kind of computing systems.

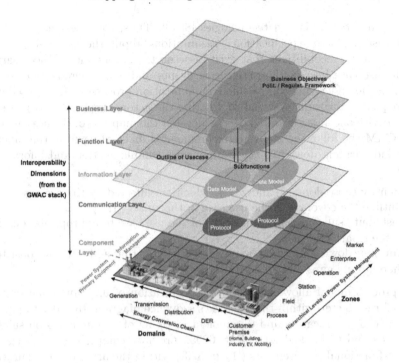

**Fig. 1.** Smart Grid Architecture Model (SGAM, [6])

For the development of interoperable Smart Grid architectures the component layer is extended by additional interoperability layers derived from the high-level categorization approach developed by the GridWise Architecture Council [7] (GWAC stack, see Figure 1). Its layers comprise a vertical cross-section of the degrees of interoperation necessary to enable various interactions and transactions within a Smart Grid. Basic functionality (e.g. interaction with field equipment, transcoding and transmitting data) is confined to the lower component and communication layers. Standards for data and information modeling and exchange are defined on the information layer while the top functional and business layers deal with business functionality. As the functions, capabilities and participating actors increase in terms of complexity, sophistication and number, respectively, more layers of the GWAC stack are utilized in order to achieve the desired interoperable results. Thus, each layer typically depends upon and is enabled (through the definition of well known interfaces) by the layers below it.

## 2.2 Methodology for Use Case Based Requirements Engineering in Energy Systems

Within this framework for a sustainable development process, which is based on the IEC PAS 62559, use cases can be formally described detailing functional and performance requirements in order to assess its applicability and (if so) appropriate standards and technologies [8] [9]. Thus, the relevant contents of a use case are actor specifications, assumptions about the use case scenario, contracts and preconditions that exist between the actors and are necessary to set the use case into motion, triggering events and a numbered list of events detailing the underlying control and communication processes. UML-based class and sequence diagrams are used to describe the actor setup and how processes operate with one another and in what order. For the mapping of the use case onto the SGAM the following steps are recommended [6] starting from the function layer and "drilling down" to the components, communication and information layers:

1. Identify those domains and zones, which are affected by the use case
2. Outline the coverage of the use case in the smart grid plane
3. Distribute sub-functions or services of the use case to appropriate locations in the smart grid plane
4. Continue down to the next layers while verifying that all layers are consistent, there are no gaps and all entities are connected.

This process model allows the detailed specification of non-functional requirements for each step within the use case's sequence diagram. In order to provide a consistent classification and to enable a mapping of requirements on suitable standards and technologies the PAS 62559 provides comprehensive check lists for non-functional requirements ([1], p. 57ff.) within the areas of Configuration, Quality of service, Security requirements, Data management issues, Constraints and other issues. The QoS-requirements do not only allow specifying what standards and protocols to use but also support the decisions regarding technology or (organization, management) algorithms in order to achieve the necessary performance, availability of the system (acceptable downtime, recovery, backup, rollback, etc.) the frequency of data exchanges, and the necessary flexibility for changes in a future Smart Grid interconnecting a vast number of sensors and actuators.

## 3  Measures for Self-organizing Systems

To this end self-organizing systems are to achieve necessary levels of service quality in complex large-scale technical systems while at the same time increasing adaptivity and robustness of the system through distribution of (critical) control. In the recent years some measures of self-organizing properties in complex systems have been developed [10], [11], [12]:

**Autonomy:** A measure for autonomy specifies the amount of external data needed to control the system.

**Emergence:** A measure for emergence indicates whether some global structures are induced by the local interactions between the entities.

**Global state awareness:** A measure for global state awareness specifies the amount of information of each single entity about the relevant global properties, such that the entities are able to make the right decisions to fulfill the overall goal of the system.

**Target orientation:** A measure for target orientation specifies whether the overall goal of the system is satisfied.

**Adaptivity:** A measure for adaptivity specifies how good the system can react to changes in the environment, such that the overall goal is still satisfied.

**Resilience:** A measure for resilience describes how good the system can react to abnormal events in the system like malfunctioned nodes, attacks against some entities of the system or a break down of some entities.

Each of these 6 measures is time dependent and yields values in the real interval $[0, 1]$, where the value 1 means that the property is fully satisfied in the current point in time, while the value 0 means that the property is not satisfied at all in the current point in time. Therefore each measure can be seen as a map $m : T \to [0, 1]$ from the time space into the interval $[0, 1]$. Averaging over time, this measure evaluates the whole system.

### 3.1 Evaluation of QoS-Requirements of Smart Grid ICT- and Automation Systems

Some of these measures used in self-organizing systems can be effortlessly applied for the evaluation of QoS-requirements of Smart Grid ICT- and automation systems. Since an analytical evaluation of the measures usually is too complex in large systems, approximation methods are needed. An evaluation of the measures can either be done by simulations or by measurements in a real system. The results can be used for the specification of design criteria in the engineering process of Smart Grids and for the optimization of existing Smart Grids.

To achieve this goal the following steps have to be performed:

1. Specify a use case including the corresponding list of QoS-requirements that have to be fulfilled.
2. Specify a micro-level model describing the entities and their interactions.
3. For each QoS-requirement specify the corresponding measure which can be used for the evaluation of the requirement.
4. Combine the measures of 3. to get an overall measure for the evaluation of the QoS-requirements.
5. Choose an evaluation method (e.g. simulations, measurements in real systems, etc.).
6. Evaluate the measure for the Smart Grid for different system parameters.
7. Find the optimal values for the system parameters, which fulfill all requirements.

A micro-level model according to step 2 can easily be specified based on the SGAM. However, step 3 requires a semantic mapping from the IntelliGrid QoS-ontology on the afore mentioned properties.

## 4  Conclusion and Future Work

A mapping between self-organization properties and the use case requirements is needed in order to support explicit design decisions towards self-organizing systems in future Smart Grids. For selected use cases in Smart Grids we plan to evaluate different QoS requirements with the developed measures [11] [12] to extract design criteria and to optimize system parameters with respect to the specified requirements. We will report on the outcoming of this evaluation in upcoming publications.

# References

1. International Electrotechnical Commission (IEC), Publicly available specification (pas) 62559 intelligrid methodology for developing requirements for energy systems (2008)
2. Rohjans, S., Daenekas, C., Uslar, M.: Requirements for smart grid ict architectures. In: 3rd IEEE PES Innovative Smart Grid Technologies (ISGT) Europe Conference (2012)
3. European Commission, Smart Grid Mandate - Standardization Mandate to European Standardisation Organisations (ESOs) to support European Smart Grid deployment. M/490 EN (2011)
4. Bruinenberg, J., Colton, L., Darmois, E., Dorn, J., Doyle, J., Elloumi, O., Englert, H., Forbes, R., Heiles, J., Hermans, P., Kuhnert, J., Rumph, F.J., Uslar, M., Wetterwald, P.: Smart grid coordination group technical report reference architecture for the smart grid version 1.0 (draft) (March 3, 2012). tech. rep. (2012)
5. ISO/IEC/IEEE, Systems and software engineering – architecture description, ISO/IEC/IEEE 42010:2011(E) (Revision of ISO/IEC 42010:2007 and IEEE Std 1471-2000), vol. 1, pp. 1–46 (2011)
6. CEN, CENELEC, ETSI, SGCP report on reference architecture for the smart grid v0.5 (for sg-cs sanity check) (January 25, 2012), tech. rep. (2012)
7. T. G. A. Council, GridWise Interoperability Context- Setting Framework (2008)
8. Uslar, M., Specht, M., Rohjans, S., Trefke, J., Gonzalez, J.M.: The Common Information Model CIM: IEC 61968/61970 and 62325 P Practical Introduction to the CIM. Springer (2012)
9. Wegmann, A., Genilloud, G.: The role of 'roles' in use case diagrams. EPFL-DSC DSC/2000/024, EPFL-DSC CH-1015 Lausanne (2000)
10. Auer, C., Wüchner, P., de Meer, H.: The degree of global-state awareness in self-organizing systems. In: Spyropoulos, T., Hummel, K.A. (eds.) IWSOS 2009. LNCS, vol. 5918, pp. 125–136. Springer, Heidelberg (2009)
11. Holzer, R., de Meer, H., Bettstetter, C.: On autonomy and emergence in self-organizing systems. In: Hummel, K.A., Sterbenz, J.P.G. (eds.) IWSOS 2008. LNCS, vol. 5343, pp. 157–169. Springer, Heidelberg (2008)
12. Holzer, R., de Meer, H.: Quantitative Modeling of Self-Organizing Properties. In: Spyropoulos, T., Hummel, K.A. (eds.) IWSOS 2009. LNCS, vol. 5918, pp. 149–161. Springer, Heidelberg (2009)

# A Self-organising Approach for Smart Meter Communication Systems

Markus Gerhard Tauber[1], Florian Skopik[1],
Thomas Bleier[1], and David Hutchison[2]

[1] AIT, Austrian Institute of Technology
{markus.tauber,florian.skopik,thomas.bleier}@ait.ac.at
[2] Lancaster University
d.hutchison@lancaster.ac.uk

**Abstract.** Future energy grids will need to cope with a multitude of new, dynamic situations. Having sufficient information about energy usage patterns is of paramount importance for the grid to react to changing situations and to make the grid 'smart'. We present preliminary results from an investigation on whether autonomic adaptation of intervals with which individual smart meters report their meter readings can be more effective than commonly used static configurations. A small reporting interval provides close to real-time knowledge about load changes and thus gives the opportunity to balance the energy demand amongst consumers rather than 'burning' surplus capacities. On the other hand, a small interval results in a waste of processing power and bandwidth in case of customers that have rather static energy usage behaviour. Hence, an ideal interval cannot be predicted *a priori*, but needs to be adapted dynamically. We provide an analytical investigation of the effects of autonomic management of smart meter reading intervals, and we make some recommendations on how this scheme can be implemented.

## 1 Introduction

Emerging alternative forms of energy are increasingly allowing consumers to produce electricity and to feed surplus capacities back into the power grid. This will turn them from consumers into producers. Additionally, power grids will become more important for mobility as electric cars will be connected to the grid for charging their batteries. These aspects will turn traditional customers into prosumers. If, for instance, a customer, produces surplus energy via solar panels during the day, (s)he would be in the producer role. When however charging her or his car over night, (s)he would become a (heavy) consumer. Such emerging scenarios will contribute to a highly dynamic overall power grid usage which requires the traditional power grid to become smarter by adding more control – the *Smart Grid*. A smart grid utility provider needs to be able to detect over-usage or under-provision in (real-)time to manage demand by, for instance,

W. Elmenreich, F. Dressler, and V. Loreto (Eds.): IWSOS 2013, LNCS 8221, pp. 169–175, 2014.
© IFIP International Federation for Information Processing 2014

scheduling the charging time of consumers' e-cars. This would avoid 'burning off'[1] surplus capacities and hence increase sustainability of energy usage.

**Motivation.** Fine-grained information about energy consumption patterns is of paramount importance for grid providers to react to a changing environment and to maintain high sustainability by, for instance, managing demands flexibly. Today's smart meters send consumption values to the grid provider at constant intervals [5,7]. A small interval is beneficial for energy sustainability and a power grid's efficiency, as it allows fine-grained demand management [1]. This can also be directly beneficial for the customer if the provider's ability to control charging periods of her/his heavy usage appliances (e.g. an electric car battery charger), is incentivised by reduced energy prices for the consumer. However, if consumers without heavy usage appliances (i.e. those who do not exhibit a high degree of variability in energy consumption), frequently report their energy usage, they reduce the benefit of small intervals as unnecessarily gathered monitoring data increases the overall operational load on the grid's ICT infrastructure. Uniformly applied intervals may not be sufficient as energy usage patterns and energy requirements vary from household to household, and over time.

**Secondary Effects: Sustainability vs. Privacy Trade-Off.** Furthermore, a small interval threatens individual customer's privacy – especially if it allows the derivation of behaviour patterns [5] from energy consumption readings. Thus, depending on the situation, a small interval may require privacy to be traded off against sustainable energy usage. We are able to provide some information to prepare further investigations even though this concern is outside the scope of this paper.

**Research Contributions.** As outlined above, different situations can be identified in which various grid aspects depend on the flexible usage patterns determined by the user base. We pick the reporting interval as first representative example to investigate how to apply more self-* properties to the smart-grid. For the considered case we identify situations where small reporting intervals are beneficial with respect to sustainability whereas large intervals can be disadvantageous. The contrary view applies with respect to efficiency in processing the reported data. Thus an ideal interval cannot be predicted *a priori* as it depends on the variability of an individual's energy usage over time. Autonomic management [2] is a (policy-driven) approach to adapt the metering report interval for *individual smart meters* in response to different usage patterns and requirements, in order to improve the grid's *overall efficiency*, in contrast to contemporary statically configured systems. Eventually, this will turn the smart grid into a self-organising system. As a direct effect, complexity is moved from a central point of computation (at the utility provider) to the individual smart meters, which increases efficiency in data processing and makes the system as a whole more scalable, more resilient and has a positive impact on the above mentioned secondary effects.

---

[1] Where the term 'burning off' energy in this case refers to the losses due to reduced degree of efficiency when employing battery buffers or pumped storage hydro power stations. See: Energy storage - Packing some power. The Economist. 2011-03-03.

**Structure of This Paper.** We provide some information on background and related work in Section 2, followed by an analysis of the problem and our approach towards autonomic management in Section 3. We report on a preliminary evaluation of our approach in Section 4, and provide some concluding remarks and outlook on future work in Section 5.

## 2   Background and Related Work

Smart Grids in general represent a popular research topic; however, neither our primary focus, which is to improve efficiency by autonomic interval adaptation nor our secondary topic of interest, which is the (autonomic) management of the trade-off between sustainability and privacy via reporting interval adaptation, is well investigated. For instance, current state of the art regarding meter reading reporting involves a static configuration and only mentions different statically configured intervals [5], but no autonomic adaptation of those is being discussed. [5] mentions that *Canada supports meter readings at 5 to 60 minute intervals* and that *the next generation of smart meters will reduce these time intervals to one minute or less.* [7]. Other existing work [9] has been conducted to apply autonomic management to multiple domains. This also includes the adaptation of house keeping intervals in order to improve routing overlays efficiency. Despite being in a different field it shares some similarities to the approach we propose.

With respect to secondary effects of our approach, threats to and vulnerabilities of smart metering systems are widely discussed topics [4,3,10]. While communication security [3] is widely studied, the aspects of privacy and potential threats [5] to customers through smart meter data exploitation are not fully covered up till now [8]. An important first step towards a privacy-enabled smart grid has been made by NIST [6], when defining problems related to privacy protection and legal constraints.

## 3   Problem Analysis and Approach

### 3.1   Scenario

Based on the demand management use case introduced in Section 1, we model the ideal frequency with which meter readings are being reported in terms of the dependency of the fluctuation of power usage over time, (i.e. in energy usage). If the power consumption and its fluctuation over time are high it is beneficial for the provider if meter readings are sent at small intervals. This is, however, only applicable if power usage of the individual customer exhibits some degree of fluctuation and intensity. Autonomic management is an approach to control systems in the presence of changing situations and requirements.

### 3.2   Approach

Autonomic management approaches in general are based on the autonomic management cycle. This comprises a monitoring, analysis, planning and execution phase. During a monitoring phase relevant events are captured, of which metrics are derived in an analysis phase. Based on these metrics a policy determines how the system is modified in a planning and an execution phase.

*At a high level:* Our autonomic management mechanism is intended to operate on each smart meter in a grid individually, requiring local data only in order to achieve an overall benefit. It is designed to detect cases when reporting effort is being wasted, and to increase the current reporting interval accordingly. Conversely, it decreases the interval in situations when a higher reporting rate is appropriate. A high variability within reported (aggregated) values suggests a decrease of the interval which other wise could be increased, to reduce unnecessary reporting activities. The magnitude of change for each of these interval adaptations (increase/decrease) is determined by our autonomic manager's policy. During each planning phase, the policy considers metric values derived from events received during the current autonomic cycle. These events are based solely on locally gathered data, thus no additional network traffic is generated by the autonomic manager.

*In detail this means that:* In the monitoring phase energy consumption is measured in (close to) real-time, and such an individual measurement is referred to as *Raw Energy Measurement (REM)*. A number of such REM values will be measured, maintained and aggregated. We refer to the aggregate values as *Aggregated Raw Measurements (ARM)*; the latter values are sent in smart meter reports. The metric we consider as an appropriate measure for variability is the standard deviation of ARM values ($\sigma_{ARM}$). Based on $\sigma_{ARM}$ the policy determines the proportion $P$ by which the current interval should be decreased. In our preliminary investigation here we define a threshold $t$ after which we consider the variability (i.e. $\sigma_{ARM}$) as high enough for decreasing the reporting interval. The new interval is then calculated as:

$$new\ interval = current\ interval \times (1 - P) \tag{1}$$

The proportion of change $P$ lies between zero and one, and is calculated as:

$$P = 1 - \frac{1}{\frac{metric - ideal}{k} + 1} \tag{2}$$

where *metric* denotes $\sigma_{ARM}$ and *ideal* is zero in our case. $k$ is a positive constant that controls the rate of change of $P$ with respect to the difference between the metric value and its ideal value. The higher the value of $k$, the lower the resulting proportion of change, and hence the slower the resulting response by the manager. $k$ can be used to consider consumers' reporting preferences in the smart meter configuration. Further, we define that, if $\sigma_{ARM}$ is smaller than the variability threshold we increment the current value by 10 (arbitrarily chosen). We constrain ourselves here to values between 1 seconds (the lowest possible reporting interval) and 1 hour (the maximum interval [7] – see Section 2).

## 4   Evaluation and Results

We have evaluated our autonomic management approach based on usage patterns derived from [6]. These show the energy consumption of a number of appliances over a normal day. We reproduce this as shown in Figure 1, which also represents measurements at the smallest possible interval (i.e. 1 sec.). The exhibited usage pattern represents an *average user* into which we factor in an e-car battery[2] to represent a regular use case with phases of *heavy usage* (afternoon/evening) and *light usage* (nights). Figure 1 shows the energy usage over the elapsed time during a day when a static energy reading interval (1 sec) is defined and also shows how the reporting interval is autonomically adapted, based on our approach (as outlined in Section 3). We choose a number of values for the policy parameters $t$ (threshold) and $k$ (adaptation rate control), the parameter values are given in the individual plots. We also configure our policy

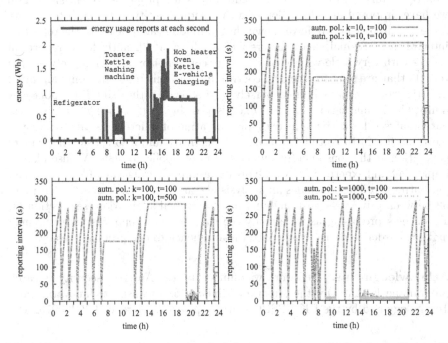

**Fig. 1.** Raw energy usage over time, at the left top and autonomically adapted intervals

to only consider a sample size of the latest 30 *ARM* values to compute $\sigma_{ARM}$. The number of values determines how much outliers may be compensated for and the semi-arbitrarily chosen number (based on test runs of our harness) was considered sufficient for this initial demonstration. We leave an investigation on ideal sample sizes for future work as this would be beyond the scope of this work.

The presented results show that our simplistic autonomic manager detects phases of little variability and increases the reporting interval accordingly until a peak – this suggests potential for improvement with respect to how aggressively the manager deals with variability and high intervals. We also see that the configuration with high $k$ values (1000) reacted most desirable by keeping the

---

[2] http://www.pluginrecharge.com/p/calculator-how-long-to-charge-ev.html

interval low when fluctuations occurred. Only little difference can be observed between the effects due to $t$. However, an holistic analysis (e.g. gradient decent) for all above mentioned policy parameters is required as next step in future work.

## 5 Conclusion

As outlined in this paper, the smart-grid will have to be increasingly flexible to cope with varying usage patters and hence the identification of aspects which can be managed in an autonomic manner is an important step to improve the smart-grid further. In a first step to add some self-* properties to the smart grid we have proposed an adaptation of an individual entity of the grid to achieve some overall benefit. We have shown that autonomic adaptation of the reporting interval in individual smart meters will result in significantly fewer reports in phases with very little variability of energy consumption behaviour between the reports that are send to a central control unit. As this represents only an initial investigation, we have limited ourselves to show a rather simplistic policy. A multitude of adaptations of our policy is possible, e.g. reducing the historical data analysed for deriving metrics, or evaluating of different parameters (e.g. $k$ see Equation 2) to consider consumer preferences, and also considering energy generation at the consumer/prosumer side in our policy design. We also plan to implement and experimentally evaluate our approach using available tools, as e.g. [9]. Future work also includes an analysis of other self-* aspects, e.g., the trade-off between privacy vs. sustainability due to interval adaptation (see Section 1). This seems promising as we already see that adapted reporting intervals may make it harder to derive usage patterns and hence to compromise privacy.

**Acknowledgments.** This work was funded by the Austrian security research programme KIRAS, by the Austrian Ministry for Transport, Innovation and Technology and by the UK EPSRC Current project (reference EP/I00016X).

## References

1. Ibars, C., Navarro, M., Giupponi, L.: Distributed demand management in smart grid with a congestion game. In: SmartGridComm, pp. 495–500 (October 2010)
2. Kephart, J.O., Chess, D.M.: The Vision of Autonomic Computing. IEEE Computer 36(1), 41–50 (2003)
3. Khurana, H., Hadley, M., Lu, N., Frincke, D.A.: Smart-grid security issues. IEEE Security & Privacy 8(1), 81–85 (2010)
4. Metke, A.R., Ekl, R.L.: Security technology for smart grid networks. IEEE Transactions on Smart Grid 1(1), 99–107 (2010)
5. Molina-Markham, A., Shenoy, P., Fu, K., Cecchet, E., Irwin, D.: Private memoirs of a smart meter. In: ACM Build Sys, pp. 61–66 (2010)
6. NIST: Nistir 7628: Guidelines for smart grid cyber security, Vol. 2, privacy and the smart grid. Tech. rep. (2010)
7. Sensus: Flexnet ami system, http://sensus.com/

8. Skopik, F.: Security is not enough! on privacy challenges in smart grids. Int'l J. of Smart Grid and Clean Energy 1(1), 7–14 (2012)
9. Tauber, M., et al.: Self-adaptation applied to peer-set maintenance in chord via a generic autonomic management framework. In: SASOW, pp. 9–16 (2010)
10. Wei, D., et al.: An integrated security system of protecting smart grid against cyber attacks. In: Innov. Smart Grid Tech., pp. 1–7 (2010)

# Unifying Sensor Fault Detection with Energy Conservation

Lei Fang and Simon Dobson

School of Computer Science, University of St Andrews UK
lf28@st-andrews.ac.uk

**Abstract** Wireless sensor networks are attracting increasing interest but suffer from severe challenges such as power constraints and low data reliability. Sensors are often energy-hungry and cannot operate over the long term, and the data they gather are frequently erroneous in complex ways. The two problems are linked, but existing work typically treats them independently: in this paper we consider both side-by-side, and propose a self-organising solution for model-based data collection that reduces errors and communications in a unified fashion.

## 1 Introduction

Wireless Sensor Network (WSN) applications have been attracting growing interest from both academia and industry. For environmental monitoring, distributed nodes are deployed to sense the environment and return sampled data to the sink, typically at a regular frequency. However, sensors are energy hungry. Various energy-saving techniques [1,2] have been proposed, such as sleep/wakeup protocols, model-based data collection, and adaptive sampling.

Another, related problem is the low reliability of data gathered by sensors. A substantial portion of the data gathered in real monitoring applications is actually faulty [3]. A data-fault can be defined as reading entries, sampled and reported by the sensor nodes, deviating from the true representation of the physical phenomenon to be measured. For example, many data series, such as [4], have been found to be faulty. Deployed WSNs are expected to operate in remote areas for long periods of time without human intervention: manual fault filtering is not an option. To increase data reliability, automatic data fault detection methods become imperative. However, most existing work adopts a centralised and off-line approach to checking the validity of data, which does not validate the sensed data until it has been collected. Recent work [5] has proposed a framework to increase sensor data reliability in a real time.

However, the two domains – energy and error – are typically excluded from each other: sensor fault detection has always been separated from sensor data collection techniques. Existing sensor fault detection methods do not care how the data is collected and whether the collection is energy-efficient or not. Conversely, energy-efficient data collection methods usually assume the validity of sensor data and so do not incorporate sensor fault detection into their collection

W. Elmenreich, F. Dressler, and V. Loreto (Eds.): IWSOS 2013, LNCS 8221, pp. 176–181, 2014.

method. In reality, these two problems both coexist and co-relate. To achieve energy-efficient, minimum-error data collection, we require a solution combining fault detection and energy-aware data collection.

In this paper, we propose an initial framework addressing both features in a self-organising fashion. We use only local data and communications to learn and autonomously maintain a model of the data expected to be observed by the network, and then uses this model both to pre-filter data to remove probable errors and to reduce the amount of raw data exchanged to reduce energy usage.

Section 2 presents our framework, which is evaluated in section 3 by means of some preliminary simulation results. Section 4 concludes with some directions for future work.

## 2   A Proposed Framework

In our proposed framework, models describing the expected behaviour of the sensed phenomena are maintained symmetrically in both network nodes and sink(s). The model is used to predict the measurements sampled by the nodes within some error bounds. Essentially when a data entry sensed is within a predefined error-band ($\epsilon$) of the value calculated by the model, the data entry is exempt from sending to the sink. However, a disagreement between the data and the model may either mean the model is stale, making the model no longer fits the changing phenomenon, or the data is faulty, resulting in it deviating from the model. Therefore, a fault-detection technique is needed to filter out faulty data from "normal" readings so that necessary updates can take place.

The whole life cycle of the framework can be described as follows. The algorithm starts with a learning phase, in which the data model (we use ARIMA model in this work, which is presented in detail in 2.1) and a spatial model (Algorithm 1) is learnt and synced between nodes and sink. Also the fault detection model, presented in detail in 2.2, is also constructed in this phase. Following the learning phase, a error-aware data collection phase starts. In each remote node, it compares data entries sensed against the model. When a disagreement occurs, the data entry will go through the fault detector first, and different actions will be taken according to the result. If the result is positive, i.e. faulty, the data entry will be ignored; otherwise, it will be sent to the sink. The whole process will go back to the learning phase again, when a large number of data entries have to be sent.

Another way to look at this approach is that the WSN develops a model of the data it is observing in a manner akin to what a scientist does when evaluating a data set – but performs this analysis automatically and on-the-fly as data is being collected. The data being collected is therefore being used to maintain and refine the model rather than being treated as uninterpreted by the network, with the model being updated dynamically to reflect the data. While not appropriate for all domains, we believe that this approach is well-suited to systems with missions to detect anomalies, systematic or other deviations from an expected observational baseline, and this class of systems has wide applicability in domains including environmental monitoring and human safety.

## 2.1  The ARIMA Model

The ARIMA model is widely used for univariate time series modelling, and has been used [2] to form an energy-efficient information-collection algorithm.

ARIMA consists of three terms: the *auto-regressive* (AR), the *moving average* (MA), and an optional *integration* term (I). The AR term is a linear regression of the current data value against prior historical data entries. The MA term captures the influence of random "shocks" to the model. The I term makes the process stationary by differencing the original data series. A first-order differencing of data series $X_t$, for example, is defined as $X_t^1 = X_t - X_{t-1}$. We use the notation ARIMA$(p, d, q)$ to define a data series modelled by a ARIMA model with $p$-lagged historic data entries, $q$ most recent random shocks, and $d$-order difference:

$$X_t^d = \phi_1 \times X_{t-1}^d + \ldots + \phi_p \times X_{t-p}^d + \varepsilon_t + \theta_1 \times \varepsilon_{t-1} + \ldots + \theta_q \times \varepsilon_{t-q} \qquad (1)$$

In the learning phase, the parameter set, including $\phi_1, \ldots, \phi_p; \theta_1, \ldots, \theta_q$, is learnt from the learning data by either conditional least squares method or maximum likelihood method. However, if the model include only AR terms, then ordinary least square method can be used, which means the whole computation can be placed in network. To sync the model, only an array of floating numbers are required to send.

## 2.2  Data Fault Detection

In this work, in order to place the detector into resource-constrained sensors, we use a simple but effective method to filter out erroneous readings. Spatial correlation is exploited to further validate an observation. To quantify the spatial correlation, each node $i$ maintains a $2 \times |nbr_i|$ spatial matrix, where $nbr_i$ denotes the set of neighbouring nodes of $i$. The spatial matrix simply stores the maximum and minimum difference in observation between node $i$ and its neighbours[1]. We use a voting mechanism to validate a suspect data entry. When a data entry does not match the statistical model, the hosting node will send the data to its neighbours. Upon receiving a validation request, each node will consult the spatial matrix and send back a boolean result accordingly. The final decision will be made according to the results solicited from the neighbours. The data will be marked as faulty if a larger number of neighbours believe so. The spatial correlation between nodes is assumed to be stable, meaning that the spatial matrix remains valid during the whole data collection process, which is a common (although not unassailable) assumption [5].

## 3    Preliminary Evaluation

We evaluate our proposed framework by simulation over an accepted benchmark real-world data set [4], collected by 54 sensors in Intel's Berkeley Laboratory.

---

[1] We assume for simplicity that all data is represented by a continuous range of values. Categorical data requires a slightly different treatment.

```
input   : nbr_i, x_t^i
output : spatialMatrix

initialise spatialMatrix of size 2 * |nbr_i| ;
for data series d_t in nbr_i do
    spatialMatrix(0, i) ← max(|d_t − x_t^i|) ;
    spatialMatrix(1, i) ← min(|d_t − x_t^i|) ;
end
```

**Algorithm 1.** Learning Spatial Model Algorithm

We examine both the fault-detection accuracy and energy-saving features respectively using our time-series and spatial correlation models.

### 3.1 Fault Model

Injecting artificial faults into real data set is a common approach to measure the accuracy of a detector [3]. We consider four particular kinds of faults: short, constant, noise, and drift. The parameters for the fault modelling, which are selected based on existing literatures [3] [5], are provided as well.

**SHORT.** Data $d_{i,t}$ is replaced by $d_{i,t} + d_{i,t} * f$ where $f$ is a random multiplicative factor. $f$ ranging from 0.1 to 10.0 is used here.

**CONSTANT.** Data $d_{i,t}$ for $t \in [t_s, t_e]$ is replaced by some random constant $c$, and $c$ is randomly selected from 30 to 999.

**NOISE.** Data $d_{i,t}$ for $t \in [t_s, t_e]$ is replaced by $d_{i,t} + x$, where $x$ is a Gaussian random variable, whose distribution is $N(0, \sigma^2)$, and $\sigma$ is set to be 2.

**DRIFT.** Data $d_{i,t}$ for $t \in [t_s, t_e]$ is replaced by $d_{i,t} + a^{t+1-t_s}$, where $a > 1$. We use natural number $e$ here.

### 3.2 Results

Table 1 shows the behaviour of the model to the different injected faults as described in the previous section. ARIMA(4,2,0) model is used here with the first 100 data entries as learning data. As there is no MA terms, the whole learning process is tractable to be carried out in remote sensor mote. The error-band, $\epsilon$, is set 0.4 Celsius degree.

It is clear that the proposed solution has high classification accuracy while keeping false alarm rate low. Note that although there is no specific rule to each fault type, for example constant or drift, the solution still exhibits good detection performance. It is also interesting to see that the false alarms rate is zero for all four cases. This is mainly because of the stationarity of the spatial correlation, making neighbouring nodes become good reference models for error detection.

In assessing the energy efficiency, in the interests of space we only list the performance of one sensor, Sensor 1 from the data set [4]. ARIMA(4, 2, 0) is still used. Figure 1 shows both actual data series sampled at a remote node and the

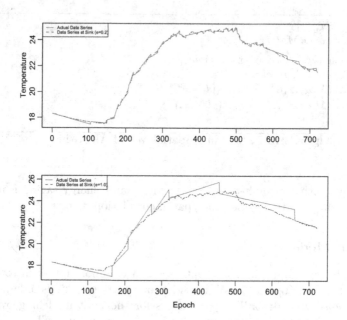

**Fig. 1.** Actual *versus* restored data at the sink node with different error-band (up ($\epsilon = 0.2\,^{\circ}C$), bottom ($\epsilon = 1.0\,^{\circ}C$))

**Table 1.** Fault Detection Accuracy

|  | SHORT | CONSTANT | NOISE | DRIFT |
|---|---|---|---|---|
| True Positive | 99.5% | 100% | 83% | 100% |
| False Positive | 0.0% | 0.0% | 0.0% | 0.0% |

data series restored at the sink as error-band is set 0.2 and 1.0 Celsius degree. It is clear that both data series restored by the ARIMA model closely follows the actual data series but at different granularities. It is obvious that the data from upper graph fits the actual data better. The mean squared error is 0.0068 and 0.2053 respectively. However the accuracy comes at the cost of higher energy consumption: only 76% of the data are exempted from sending back to the sink, while the bottom one saves 95%. But both of them save great amount of energy from sending data back to sink. Note the overall energy saving also depends on the size of data faults. As more data faults present, more extra efforts are paid to clean them out by local communication.

In general, there is a trade-off between energy saving and data integrity and granularity. Better restored data and data integrity is achieved at the price of relative frequent communication, or higher energy cost. The trade-off can be set by the error-band parameter. Users can choose its value according to specific application requirements.

# 4    Conclusion

We have presented a framework which features couples fault detection and energy efficiency in data collection. We learn and maintain statistical models of the expected data stream and use these models to reduce communications and improve bad-point rejection. Potentially they may also be used to answer queries at the sink without recourse to the WSN.

We believe that our preliminary results show that these two significant challenges to WSN deployments can be addressed using a unified approach, in a way that maintains the statistical and properties of the data collected. We believe that this suggests a practical approach to further improving WSN longevity in the field.

The work presented here remains proof-of-concept. We are currently implementing our framework on real sensors to perform a more thorough evaluation on the method, and also to test its feasibility in a real-world deployment. Other data modelling and fault detection methods are also being investigated, and we hypothesis that a range of techniques can be incorporated into the same framework to target different classes of sensor network mission. We would also like to move beyond model-based data collection methods to investigate other energy-saving techniques that have the potential to integrate further with the fault detector.

**Acknowledgements.** Lei Fang is supported by a studentship from Scottish Informatics and Computer Science Alliance (SICSA).

# References

1. Schurgers, C., Tsiatsis, V., Srivastava, M.: STEM: Topology management for energy efficient sensor networks. In: Proceedings of the 3rd IEEE Aerospace Conference (2002)
2. Liu, C., Wu, K., Tsao, M.: Energy efficient information collection with the ARIMA model in wireless sensor networks. In: Proceedings of the IEEE Global Telecommunications Conference, GLOBECOM 2005 (2005)
3. Sharma, A., Golubchik, L., Govindan, R.: Sensor faults: Detection methods and prevalence in real-world datasets. ACM Transactions on Sensor Networks 6(3), 23–33 (2010)
4. INTEL: Intel Berkeley Laboratory sensor data set (2004),
   http://db.csail.mit.edu/labdata/labdata.html
5. Kamal, A.R.M., Bleakley, C., Dobson, S.A.: Packet-Level Attestation: a framework for in-network sensor-data reliability. ACM Transactions on Sensor Networks (to appear)

# Towards a Distributed, Self-organising
# Approach to Malware Detection
# in Cloud Computing

Michael R. Watson, Noor-ul-Hassan Shirazi, Angelos K. Marnerides,
Andreas Mauthe, and David Hutchison

School of Computing and Communications, Lancaster University
Lancaster, UK, LA1 4WA
{m.watson1,n.shirazi,a.marnerides2,a.mauthe,d.hutchison}@lancs.ac.uk

**Abstract.** Cloud computing is an increasingly popular platform for
both industry and consumers. The cloud presents a number of unique
security issues, such as a high level of distribution and system homogeneity, which require special consideration. In this paper we introduce
a resilience architecture consisting of a collection of self-organising resilience managers distributed within the infrastructure of a cloud. More
specifically we illustrate the applicability of our proposed architecture
under the scenario of malware detection. We describe our multi-layered
solution at the hypervisor level of the cloud nodes and consider how
malware detection can be distributed to each node.

## 1 Introduction

Cloud environments are in general made up of a number of physical machines
hosting multiple virtual machines (VMs) that provide the resources for the
cloud's services. The datacentre has an internal network and is connected through
one or more ingress/egress routers to the wider Internet. In order to provide resilience within a cloud environment it is necessary to observe and analyse both
system and network behaviour, and to take remedial action in case of any detected anomalies.

Detection in this scenario has to happen at various points throughout the
cloud; resilience managers need to exchange information and co-ordinate a reaction to any observed anomalies. Since cloud environments are highly distributed
structures with no prescribed hierarchy or fixed configuration resilience managers
need to have the ability to flexibly organise themselves taking into account ar-
chitectural considerations as well as system state. Each resilience manager needs
to be a self-organising entity within a larger resilience management framework,
which acts autonomously but in a coordinated manner in order to maintain
overall system operability.

In this paper we present the architecture of a Cloud Resilience Manager
(CRM) and the overall architecture arising from a network of CRMs under the

W. Elmenreich, F. Dressler, and V. Loreto (Eds.): IWSOS 2013, LNCS 8221, pp. 182–185, 2014.
© IFIP International Federation for Information Processing 2014

same conceptual autonomic properties followed by previous work [1, 2]. Overall, we discuss the self-organising aspect of each element and how each CRM interacts to form the overall resilience framework.

## 2  System Architecture

The overall system architecture can be seen in Figure 1 with $A$ representing a single hardware node in the cloud. For simplicity only three nodes are shown and the network connections between each node have been omitted. Each node has a hypervisor, a host VM (or dom0 under Xen terminology[3]) and a number of guest VMs. Within the host VM of each node there is a dedicated CRM which comprises one part of the wider detection system. The internal structure of the CRM is shown in more detail by $B$ in Figure 1.

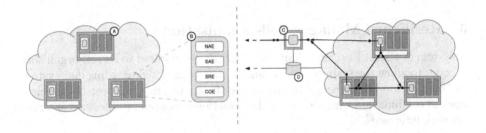

**Fig. 1.** An overview of the detection system architecture

The software components within $B$ are, in order: the Network Analysis Engine (NAE), the System Analysis Engine (SAE), the System Resilience Engine (SRE) and the Coordination and Organisation Engine (COE).

The role of the SAE and NAE components is to perform local malware detection based on the information obtained through introspection of VMs and local network traffic capture respectively. In the NAE observations from lower layers, such as the transport layer, are correlated with observations from higher layers, such as the application layer, in order to attribute anomalies across layers to the same source. In the SAE features such as memory utilisation are extracted from the processes within each VM using introspection[4] and analysed using anomaly detection techniques.

The SRE component is in charge of protection and remediation actions based on the output from the NAE and SAE. The SRE is designed to alleviate the COE of any responsibility regarding system state due to its potentially heavy workload.

Finally, the COE component coordinates and disseminates information between other instances and, in parallel, controls the components within its own node. The COE is required to correlate NAE and SAE outputs by mapping statistical anomalies found in the network to end-system state as reported by the SAE. An example of this is the identification of protocols and ports used by anomalous traffic and the attribution of these to a particular process executing within a guest VM. In this way it is possible to attribute anomalies at disparate locations in the architecture to a single threat.

In addition to node level resilience, the detection system is capable of gathering and analysing data at the network level through the deployment of network CRMs as shown by $C$ in Figure 1. $D$ in the figure represents an ingress/egress router of the cloud; the monitoring system is directly connected to router $D$ and as such can gather features from all traffic passing through it.

Self-organization in a system of CRMs is achieved through the dissemination and exchange of meaningful information with respect to the system and network activities of each VM, as well as with the router(s) that connect the datacenter network to the Internet. In practice, and as depicted in Fig. 1, there are various system/network interfaces that act as information dispatch points[1] in order to allow efficient event dissemination.

## 3    Resilience Manager Self-organisation

A system of Cloud Resilience Managers (CRMs) is required to be self-organising in order to allow the system to make autonomous decisions regarding end-system and overall cloud resilience. This is achieved through an internal peer-to-peer network in combination with hierarchical interactions between internal and external network interfaces.

### 3.1    Network Architecture

In Figure 1 the system architecture is shown as consisting of two levels of communication. The interfaces between the CRMs within the cloud correspond to an internal peer-to-peer network where peers can perform push/pull actions with other peers. The interfaces between system level CRMs and those in the network correspond to external interfaces that only allow push actions, from low level CRMs (i.e. system level) to higher level CRMs (i.e. network level).

Due to the hierarchical nature of the system the information sent to $C$ is under a filtered, event-based format resulting from the malware analysis and detection achieved internally by the SAE and NAE components. Thus, the COE in $C$ will only receive meaningful input from a remote COE such that it is able to correlate the VM-level anomalous activity with the traffic it captures on the ingress/egress router(s) $(D)$. For instance if a destination within the cloud infrastructure is locally flagged as suspicious by its CRM, traffic to that destination can be thoroughly analysed by the detection component in $C$, as in [2].

### 3.2    Peer Communication

Peer-to-peer communication of data between CRMs enables each individual CRM to make local decisions which are influenced by the activity experienced on remote cloud nodes. This direct communication between peers results in the ability of the CRMs to exchange information with respect to the current health of their respective virtual environments.

The peer network itself is a simple message exchange system, whereby healthy peers advertise their presence in the network. Peers experiencing anomalous behaviour exchange the type of anomaly and pertinent information on how this will affect other peers. The other COEs in the cloud will receive this data and take action by invoking their local SRE.

The benefits of this information exchange versus a centralised system are the ability to notify vulnerable systems in a single exchange, and the ability to reduce the amount of data flowing over communication channels. In a centralised detection system it would be necessary to export all data relating to the state of every physical machine in the cloud to a single point. This scenario puts a higher demand on network links than the solution proposed in this paper. Moreover, a single point of analysis reduces the resilience of the cloud due to a single point of failure. This fact, coupled with the inherently distributed nature of clouds indicates that self-organisation is a better fit architecturally.

## 4   Conclusion

Cloud environments present unique challenges in terms of security and resilience. These challenges need to be confronted through the synergistic analysis of both system and network-level properties in order to more effectively utilise available information. In this paper we have proposed a solution to these challenges by introducing the concept of a Cloud Resilience Manager (CRM) which combines self-organisation with a distributed approach to detection.

**Acknowledgments.** The authors would like to thank Fadi El-Moussa and Ben Azvine of BT Research for their contribution to this work. This work is sponsored by the IU-ATC (EPSRC grant number EP/J016675/1) and the EU FP7 SECCRIT project (grant agreement no. 312758).

## References

[1] Marnerides, A.K., Pezaros, D.P., Hutchison, D.: Detection and mitigation of abnormal traffic behaviour in autonomic networked environments. In: Proceedings of ACM SIGCOMM CoNEXT Conference 2008 (2008)
[2] Marnerides, A., Pezaros, D., Hutchison, D.: Autonomic diagnosis of Anomalous network traffic. In: Proceedings of IEEE WoWMoM 2010 (2010)
[3] Citrix Systems, Inc., Xen, http://www.xen.org/
[4] Payne, B.D.: LibVMI,
http://code.google.com/p/vmitools/wiki/LibVMIIntroduction

# Demand Response by Decentralized Device Control Based on Voltage Level*

Wilfried Elmenreich[1,2] and Stefan Schuster[1]

[1] Dep. of Informatics and Mathematics
University of Passau, Germany
[2] Netw. and Emb. Systems/Lakeside Labs
Alpen-Adria-Universität Klagenfurt, Austria
wilfried.elmenreich@uni-klu.ac.at

**Abstract.** This paper introduces a distributed, self-organizing approach to load control based on voltage measurement. A local voltage measurement defines a Level of Service (LoS), which is balanced with the neighboring households in order to avoid extreme restrictions of energy use in a single household. The approach distinguishes four criticality classes for devices which suspend themselves at specific LoS thresholds.

## 1 Introduction

The amount of alternative energy sources such as photovoltaic systems or wind power will increase significantly in future. Renewable energy sources, however, typically rely on the weather and thus lead to variable energy production which is hard to manage [1]. In this work, we concentrate on how to handle power under-supply, i.e., the situation where power demand exceeds possible production capability. Based on the ColorPower architecture described in [2], we propose a decentralized control mechanism to temporarily turn off devices dependent on their criticality. Although in [2] all modeled devices consume an equal amount of energy, we consider devices with a wider range of power consumption, targeting typical domestic application. Following [3], we organize devices into households and take effects such as connection resistivity into account.

## 2 Device Model

A user or given schedule decides when a device is needed. This is modeled by a starting time $t_{on}$ and a duration of use, $d_{on}$, which are both normally distributed with standard deviations $\sigma_{t_{on}}$ and $\sigma_{d_{on}}$, respectively. Requests for using a device are typically periodically with period $T$. As an example, the parameters for a stove could look like this: $t_{on} = 11AM$, $\sigma_{t_{on}} = 0.5$, $d_{on} = 0.75$, $\sigma_{d_{on}} = 0.3$,

---

* This work is supported by the Carinthian Economic Promotion Fund (KWF) under grant 20214/22935/34445 (Project Smart Microgrid Lab). We would like to thank Lizzie Dawes for proofreading the paper.

W. Elmenreich, F. Dressler, and V. Loreto (Eds.): IWSOS 2013, LNCS 8221, pp. 186–189, 2014.

$T = 24h$, i.e., the stove is used daily around $11AM$ for about 45 min. A device currently requested is called *requested*. As long as the energy provided is sufficient, all requested devices are turned on, i.e., they are *active*. If a device is requested but cannot be turned on because of energy shortage, it is *suspended*. Depending on their criticality, devices are classified into four categories identified by the colors green, blue, red and black based on the ideas in [2]. "Green" devices can be suspended without disturbing the user significantly and without losing vital services. An example of a green device is a washing machine programmed by the user to wash within the next few hours, but not necessarily immediately. "Blue" devices are not vital to the infrastructure, but annoy the user if they are not working, e.g., supplemental lighting systems. "Red" devices significantly affect the user, for example food in a refrigerator will spoil faster if the refrigerator is turned off. "Black" devices are the most vital services for the infrastructure that should be maintained operational as long as possible.

## 3  Device Control by Voltage Level

The voltage level can be used as a measure for the amount of energy available in the system; voltage drops if not enough energy can be supplied. The standard IEC 60038 defines valid voltage as $\pm 10\%$ of the nominal voltage.

Devices in the household can counteract a supply shortage by reducing demand upon a voltage drop. This is done by suspending devices.

Figure 1 describes a mapping of voltage levels to the four types of criticality classes or labels. A particular voltage level defines which device classes have to be turned off completely, which classes are unaffected, and which device class is partially suspended.

| V | 230V | 225V | 220V | 215V |
|---|---|---|---|---|
| LoS | [4.0 − 3.0) | [3.0 − 2.0) | [2.0 − 1.0) | [1.0 − 0.0] |
| green | s | o | o | o |
| blue | - | s | o | o |
| red | - | - | s | o |
| black | - | - | - | s |

**Fig. 1.** Exemplary mapping between labels and voltage levels. For example, if the voltage is between 220 and 225$V$, part of the blue devices are suspended (s), below 220$V$ all blue devices are turned off (o).

Figure 2 shows the possible states and state transitions of a device. If a user wants to turn on a device, the device goes into state *on-protected* (unless the voltage level does not allow at that time for using that class of devices – if this is the case, the device goes into the state *suspended-protected*). A state with addition *protected* means that the device will stay in this state for a predefined amount of time in order to avoid devices switching on and off repeatedly. Partial suspension of a device class is implemented by a probabilistic algorithm. Based on the function in Figure 2, a particular probability level $p$ defines the probability

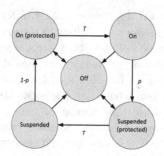

**Fig. 2.** Device control states

of a device being suspended. After the protection time $T$, a device can come back on with the reverse probability $1 - p$. After some time, this algorithm converges towards having a ratio $p$ of suspended devices.

## 4   Fairness by Synchronization

The voltage drop gives good feedback on the local grid situation, however, due to different cable lengths, there are likely to be voltage differences between neighboring houses. To ensure fairness, we require that within a small distance, e.g., from one house to another, the amount of suspended devices should not differ significantly. On the other hand, if a larger area, e.g., a whole neighborhood, experiences a voltage drop, the consumers in the area should collectively react to the problem. Therefore, we introduce a layer between measured voltage and mapping in Figure 1 named "Level of Service" (LoS). The LoS is repeatedly calculated from the local measured voltage level and an averaging process based on the neighbors' communicated LoS as follows:

1. $LocalLoS = f(Voltage)$
2. Receive LoS values from neighbors via local wireless communication
3. $LoS = \alpha \cdot average(neighbors' LoS) + (1 - \alpha)LocalLoS$
4. Broadcast LoS

The coupling factor, $\alpha$, determines how much the LoS is influenced by the situation in the neighborhood.

## 5   Simulation

For evaluation we simulate up to 10 streets, each consisting of 20 to 50 households linked together and connected to a power provider (e.g., a mid/low voltage transformer).The wireless network for synchronization is based on a CC2420 radio with a transmission range of approximately $30m$. The model is based on [3]. The model accounts for the resistance of cables between households. These resistances are derived from physical and electrical properties of typical feed cables

(street cable $150mm^2$, building connection $35mm^2$). We implement a simple energy flow model without reactive loads. The energy consumption of each household is modeled via a number of devices, which are modeled according to the device model described in Section 2 We randomly assigned up to $\pm10\%$ of the required energy for each street, causing energy shortages in some of the streets. If the total load of households in a street corresponds to higher power than provided, this results in a lower voltage at the feed and, consequently, in reduction of power consumption proportional to $(\frac{U_{nominal}-U_{drop}}{U_{nominal}})^2$ for ohmic devices. The left table in Figure 3 shows the LoS map based on the local voltage in each house. The feed is from left to right, so LoS are generally higher on the left side. There are significant differences depending on the respective feed section. The right part of Figure 3 shows the LoS after update with the synchronization algorithm with a coupling factor $\alpha = 0.5$.

| 1.9 | 1.9 | 1.9 | 1.9 | 1.8 | 1.8 | 1.8 | 1.8 | 1.8 | 1.8 |
|-----|-----|-----|-----|-----|-----|-----|-----|-----|-----|
| 4.0 | 4.0 | 4.0 | 4.0 | 4.0 | 4.0 | 4.0 | 4.0 | 4.0 | 4.0 |
| 4.0 | 4.0 | 4.0 | 4.0 | 4.0 | 4.0 | 4.0 | 4.0 | 4.0 | 4.0 |
| 2.2 | 2.1 | 2.1 | 2.1 | 2.1 | 2.0 | 2.0 | 2.0 | 2.0 | 2.0 |
| 2.1 | 2.0 | 2.0 | 2.0 | 1.9 | 1.9 | 1.9 | 1.9 | 1.9 | 1.9 |
| 3.6 | 3.6 | 3.5 | 3.5 | 3.5 | 3.5 | 3.4 | 3.4 | 3.4 | 3.4 |
| 3.2 | 3.2 | 3.1 | 3.1 | 3.1 | 3.1 | 3.0 | 3.0 | 3.0 | 3.0 |
| 4.0 | 4.0 | 4.0 | 4.0 | 4.0 | 4.0 | 4.0 | 4.0 | 4.0 | 4.0 |
| 2.8 | 2.7 | 2.7 | 2.7 | 2.7 | 2.6 | 2.6 | 2.6 | 2.6 | 2.6 |
| 4.0 | 4.0 | 4.0 | 4.0 | 4.0 | 4.0 | 4.0 | 4.0 | 4.0 | 4.0 |

| -2.5 | 2.8 | 2.6 | 2.0 | 2.0 | 2.4 | 2.6 | 2.6 | 2.5 | 2.9 |
|-----|-----|-----|-----|-----|-----|-----|-----|-----|-----|
| 3.6 | 4.0 | 3.8 | 3.4 | 3.3 | 3.5 | 3.8 | 3.7 | 3.7 | 3.8 |
| 3.4 | 3.7 | 3.7 | 3.2 | 3.4 | 3.5 | 3.7 | 3.8 | 3.8 | 3.6 |
| 2.6 | 2.7 | 2.7 | 2.1 | 2.5 | 2.6 | 2.7 | 2.7 | 2.7 | 2.7 |
| 2.5 | 2.3 | 2.5 | 2.4 | 2.2 | 2.5 | 2.6 | 2.6 | 2.6 | 2.5 |
| 3.1 | 3.5 | 3.4 | 3.2 | 3.0 | 3.2 | 3.4 | 3.4 | 3.4 | 3.4 |
| 3.1 | 3.4 | 3.2 | 3.0 | 3.0 | 3.0 | 3.2 | 3.1 | 3.3 | 3.2 |
| 3.7 | 3.7 | 3.6 | 3.5 | 3.4 | 3.5 | 3.7 | 3.6 | 3.7 | 3.5 |
| 3.3 | 3.0 | 3.0 | 3.0 | 2.8 | 2.9 | 3.0 | 2.8 | 3.1 | 3.0 |
| 3.2 | 3.7 | 3.6 | 3.5 | 3.5 | 3.5 | 3.7 | 3.7 | 3.9 | 3.9 |

**Fig. 3.** Level of Service before and after synchronization

## 6    Conclusion

We presented a distributed, self-organizing approach to load control based on voltage measurement. A local wireless network and a synchronization algorithm are used to improve fairness between neighboring consumers. The approach is suited to handle networks with supply shortages, thus is especially of interest for islanded grids with power sources from alternative energy such as photovoltaic systems or wind power. For a fine-grained demand response, each house needs smart appliances which can be controlled by the presented system. Future work will include simulations with improved appliance models and heterogeneous distributions of houses.

## References

[1] Sobe, A., Elmenreich, W.: Smart microgrids: Overview and outlook. In: Proc. of the ITG INFORMATIK Workshop on Smart Grids, Braunschweig, Germany (2012)
[2] Beal, J., Berliner, J., Hunter, K.: Fast precise distributed control for energy demand management. In: Proc. of the Sixth IEEE International Conference on Self-Adaptive and Self-Organizing Systems, Lyon, France, pp. 187–192 (2012)
[3] Okeke, I.O.: The influence of network topology on the operational performance of the low voltage grid. Master's thesis, Delft University of Technology (2012)

# Author Index